IUBS Section of Comparative Physiology and Biochemistry
1st International Congress, Liège, Belgium, August 27–31, 1984

Conference Organization

Organizing Board
R. Gilles, Chairman, Liège, Belgium
M. Gilles-Baillien and L. Bolis, Liège, Belgium/Messina, Italy

Host Society
European Society for Comparative Physiology
and Biochemistry

Under the Patronage of
The European Economic Community
The Fonds National de la Recherche Scientifique
The Ministère de l'Education Nationale et de la Culture Française
The Fondation Léon Fredericq
The University of Liège

The European Society for Comparative Physiology and Biochemistry
The American Society of Zoologists
The Canadian Society of Zoologists
The Japanese Society for General and Comparative Physiology

The Congress has been organized in relation with the 100th Anniversary of the School of Comparative Physiology and Biochemistry of the University of Liège.

The proceedings of the invited lectures to the different symposia of the congress have been gathered in five different volumes published by Springer-Verlag under the following titles:

Circulation, Respiration, and Metabolism
Current Comparative Approaches
Edited by R. Gilles (ISBN 3-540-15627-5)

Transport Processes, Iono- and Osmoregulation
Current Comparative Approaches
Edited by R. Gilles and M. Gilles-Baillien (ISBN 3-540-15628-3)

Neurobiology, Current Comparative Approaches
Edited by R. Gilles and J. Balthazart (ISBN 3-540-15480-9)

Respiratory Pigments in Animals, Relation Structure-Function
Edited by J. Lamy, J.-P. Truchot, and R. Gilles (ISBN 3-540-15629-1)

High Pressure Effects on Selected Biological Systems
Edited by A. J. R. Péqueux and R. Gilles (ISBN 3-540-15630-5)

Respiratory Pigments in Animals

Relation Structure-Function

Edited by
J. Lamy, J.-P. Truchot, and R. Gilles

With 63 Figures

Springer-Verlag
Berlin Heidelberg New York Tokyo

Professor Dr. JEAN LAMY (Scientific Editors)
Laboratory of Biochemistry
Department of Pharmaceutical Science
University of Francois-Rabelais
2, Boulevard Tonnellé
F-37032 Tours Cedex, France

Professor Dr. JEAN-PAUL TRUCHOT
Laboratory of Neurobiology and Compared Physiology
University of Bordeaux I
Place du Dr. Bertrand Peyneau
F-33120 Arcachon Cedex, France

Professor Dr. RAYMOND GILLES (Coordinating Editor)
Laboratory of Biochemistry
University of Liège
22, Quai Van Benden
B-4020 Liège, Belgium

ISBN-13: 978-3-540-15629-1   e-ISBN-13: 978-3-642-70616-5
DOI: 10.1007/978-3-642-70616-5

Library of Congress Cataloging-in-Publication Data. Main entry under title: Respiratory pigments in animals. Partial proceedings of the First International Congress of Comparative Physiology and Biochemistry organized at Liège, Belgium, in August 1984 under the auspices of the Section of Comparative Physiology and Biochemistry of the International Union of Biological Sciences. Includes Index. 1. Blood—Pigments—Structure-activity relationships—Congresses. 2.Hemocyanin—Structure-activity relationships—Congresses. 3. Hemoglobin—Structure-activity relationships—Congresses. I. Lamy, Jean, 1941-. II. Truchot, Jean-Paul, 1937-. III. Gilles, R. IV. International Congress of Comparative Physiology and Biochemistry (1st : 1984 : Liège, Belgium) V. International Union of Biological Sciences. Section of Comparative Physiology and Biochemistry. QP99.3.P5R47 1985 591.1'13 85-22075

Printing and bookbinding: Beltz Offsetdruck, Hemsbach/Bergstr.
2131/3130-543210

Foreword

This volume is one of those published from the proceedings of the invited lectures to the First International Congress of Comparative Physiology and Biochemistry I organized at Liège (Belgium) in August 1984 under the auspices of the Section of Comparative Physiology and Biochemistry of the International Union of Biological Sciences. In a general foreword to these different volumes, it seems to me appropriate to consider briefly what may be the comparative approach.

Living organisms, beyond the diversity of their morphological forms, have evolved a widespread range of basic solutions to cope with the different problems, both organismal and environmental, with which they are faced. Soon after the turn of the century, some biologists realized that these solutions can be best comprehended in the framework of a comparative approach integrating results of physiological and biochemical studies done at the organismic, cellular and molecular levels. The development of this approach amongst both physiologists and biochemists remained, however, extremely slow until recently. Physiology and biochemistry have indeed long been mainly devoted to the service of medicine, finding scope enough for their activities in the study of a few species, particularly mammals. This has tended to keep many physiologists and biochemists from the comparative approach, which demands either the widest possible survey of animal forms or an integrated knowledge of the specific adaptive features of the species considered. These particular characteristics of the comparative approach have, on the other hand, been very attractive for biologists interested in the mechanisms of evolution and environmental adaptations. This diversity of requirements of the comparative approach, at the conceptual as well as at the technological level, can easily account for the fact that it emerged only slowly amongst the other new, more rapidly growing, disciplines of the biological sciences. Although a few pioneers have been working in the field since the beginning of the century, it only started effectively in the early 1960's. 1960 was the date of the organization of the periodical "Comparative Physiology and Biochemistry" by Kerkut and Scheer and of the publication of the first volumes of the comprehensive treatise "Comparative Biochemistry" edited by Florkin and Mason. These publications can be considered as milestones in the evolution of the comparative approach. They have been followed by many others which have greatly contributed to giving the field the international status it deserved. Since the 1960's, the comparative approach has been maturing and developing more and more rapidly into the independent discipline it now is, widely recognized by the international communities of physiologists, biochemists and biologists. It is currently used as an effective tool of great help in the understanding of many research problems: biological as well as clinical, applied as well as fundamental.

The actual development of the field and the interest it arouses in a growing portion of the biological scientific community led some of us to consider the organization of an international structure, bringing together the major representative societies and groups around the world, which would aim at the general advancement and promotion of the comparative approach. This was done in 1979 with the incorporation, within the international Union of Biological Sciences, of a Section of Comparative Physiology and Biochemistry. The first International Congress of CPB, I organized in Liège with the help of a few friends and colleagues, is the first activity of this newly founded Section. In 22 symposia it gathered some 146 invited lectures given by internationally renowned scientists on all major current topics and trends in the field. The proceedings of these lectures have been collected in 5 volumes produced by Springer-Verlag, a publisher long associated with the development of CPB. The organization of the CPB Section of IUBS, its first Congress and these proceedings volumes can well be considered as milestones reflecting the international status and the maturity that the comparative approach has gained, as a recognized independent discipline, in the beginning of the 1980's, some 20 years after it was effectively launched.

Finally, I would like to consider that the selection of Liège for this first International Congress has not been simply coincidental. I thus feel that this brief foreword would not be complete without noting the privileged role Liège has played in some events associated with the development of the comparative approach. Liège had a pioneer in comparative physiology already at the end of the last century with Léon Fredericq. With Marcel Florkin, Liège had its first Professor of biochemistry and one of the founding fathers of comparative biochemistry. These two major figureheads of the comparative approach founded and developed what is actually called the Liège School of Comparative Physiology and Biochemistry, which was, at the time of the Congress, celebrating its 100th anniversary. This school provided early support to the European Society for Comparative Physiology and Biochemistry organized by Marcel Florkin and myself some years ago. The society, still headquartered in Liège, was, with the CPB division of the American Society of Zoologists, at the origin of the formation of the CPB Section of IUBS under the auspices of which this first International Congress, specifically devoted to the comparative approach, has been organized. An essential particularly of the Liège school of CPB is that its two founding fathers, scientists interested in general, basic aspects of the organization of living organisms, were also professors at the faculty of medicine. This largely contributed in Liège to avoiding the undesirable structuration of a so-called "zoophysiology" or "zoobiochemistry" independent of the rest of the field. The conditions were thus realized very early in Liège for CPB to play its key role in canalizing the necessary interactions between the general, pre-clinical or clinical and the environmental, ecological or evolutionary tendencies of physiology and biochemistry. The possibility of stimulating such interactions has served as a major guide line in the selection of the symposia and invited lectures from which these proceedings have issued.

August, 1985

R. GILLES

Preface

Evolution has exploited the coordination chemistry of transition metals in many subtle ways. Some of the most ingenious and original solutions of chemical problems are to be found in the designs of metal complexes which combine reversibly with molecular oxygen. Supposing an inorganic chemist ignorant of biochemistry had been asked which metal Nature was most likely to have used in making an oxygen-carrying protein, he would probably have excluded iron from his list of guesses in view of its liability to become irreversibly oxidised. Yet this is precisely the metal that Nature has used in nearly all species perhaps because of the ease of its change from five to six coordination and the facility with which the strength of its bond with oxygen can be tuned to physiological needs.

Combination of ferrous iron with molecular oxygen needs donation of electron density from the iron to the oxygen so that the complex acquires partial Fe^{+}-O_2^{-} character. Coordination of the iron to the free nitrogen of a porphyrin and to the imidazole of a histidine has provided just the right degree of electron donation for an iron oxygen bond strength of about one fifth of that of a carbon-carbon single bond. For optimum strength the Fe-O-O angle should be about $120°$. The partial negative charge on the oxygen allows it to form hydrogen bonds with donor groups such as histidines or glutamines. The change of the iron from five to six coordination is accompanied by a shortening of the bonds between the iron and the porphyrin nitrogens and a movement of the iron relative to the plane of the porphyrin nitrogens, from out of plane in the deoxygenated to in plane in the oxygenated state.

These properties have offered Nature several ways of tuning the oxygen affinity. It can tailor the distal heme pocket so as to vary the steric hindrance to the binding of oxygen, it can alter the number and strength of hydrogen bonds to the bound oxygen and it can design the protein so as to promote or hinder the movement of the iron into the porphyrin plane. The combination of these devices has resulted in a range of oxygen equilibrium constants that vary over at least four orders of magnitude from Ascaris hemoglobin with K_a = 145 to human hemoglobin in the T structure with K_a = 0.008. The equilibrium constant of trout IV homoglobin at acid pH may be several orders smaller still.

Hemerythrins have evolved an entirely different coordination complex, consisting of a pair of iron atoms bridged by two carboxylate ions and an oxygen atom, and coordinated to five histidines of the protein, leaving one coordination site at one of the irons open for coordination with oxygen. It seems that this iron-protein complex combines with oxygen without undergoing a significant change of structure.

The pair of copper atoms that form the active site of hemocyanin probably form a bridged dioxygen complex that links the two copper atoms at a distance of $\cong$ 3.5 Å. EXAFS shows that dissociation of oxygen is accompanied by a stereochemical change, probably consisting of a movement apart from the two copper atoms. The precise coordination of the copper atoms is not yet known. The third metal used for reversible combination with oxygen is vanadium, but the chemistry of oxygen carriers containing this metal is still unexplored.

Once evolution had solved the problem of designing transition metal complexes capable of reversible combination with oxygen, the next step essential for the evolution of fast moving animals was the development of cooperative oxygen binding. One could well imagine an oxygen carrier made up of hemes in parallel pockets of a single polypeptide, joined and geared so that movement of any one heme iron towards the porphyrin plane pushes all the others in the same direction and thus raises their oxygen affinity. Why has Nature never adopted such a design? Perhaps because such a simple sequential mechanism does not lend itself to the generation of heterotropic cooperative effects needed for regulation of the oxygen affinity and evolution of two-way respiratory carriers.

In every instance discovered so far, cooperative effects are exhibited only by oligomeric proteins in equilibrium between alternative states with different oxygen affinities; these may be either alternative quaternary structures or alternative states of aggregation. Such alternative states arise only if combination of the metal with oxygen is accompanied by a stereochemical change at the metal or at the protein surrounding it. Consequently cooperative effects are exhibited only by hemoglobin and hemocyanins but not by hemerythrins.

A further important step in evolution has been the enclosure of oxygen carriers in erythrocytes. In order to contain oxygen carriers in the blood vessels they had to have very large molecular weights. Judging by the properties of erythrocruorins and hemocyanins, such large complexes do not exhibit as much chemical versatility as the $\alpha\beta$ tetramer in erythrocytes.

The pages in this volume show how fantastically adaptable hemoglobin compared to hemocyanin and hemerythrin. Not only does its oxygen affinity vary over an enormous range, but the number of subunits per molecule and with them the free energy of cooperativity, the strength and direction of the Bohr effect, the nature and binding energy of the allosteric effectors and even the thermodynamics of the reaction with oxygen vary tremendously. Perhaps this is why animals with hemoglobin have evolved so successfully.

One of the most impressive features of the hemoglobins is the constancy of their tertiary structure, ranging all the way from leg-hemoglobins that are coded for by plant genes, to invertebrate and vertebrate hemoglobins. One would have imagined this constancy to be the result of an invariant amino acid sequence, but in fact the only invariant residues are the proximal histidine and a phenylalanine that wedges the heme into its pocket. The only feature common to all the hemoglobins is a set of internal sites which are invariably occupied by non-polar amino acid residues ensuring the exclusion of polar residues from the interior. This appears to have been sufficient to preserve the vitally important geometry of the heme pocket, but allowed the structures to vary in detail. The angles between helical segments differ by up to $20°$ and the points of contact between them by up to 7 Å. Many different combinations of side-chains are found to produce helix interfaces that are comparably well packed, as if the tertiary structure had been preserved by a patchwork of improvisations.

X-ray analysis has shown that despite the great complexity of hemoglobin, the stereo-chemical mechanism of the cooperative effects is comparatively simple. The deoxy or T-structure is constrained by hydrogen bonds between and within the subunits which oppose the changes of tertiary structure required for combination with oxygen, while in the oxy or R-structure these bonds are absent. Some of these bonds are between anionic groups with pK_a's below 4 and cationic groups with pK_a's between 6 and 8. The closure of these hydrogen bonds leads to a rise and the rupture to a drop in the pK_a's of the cationic groups. These changes in pK_a's are responsible for the Bohr effect. There have been reports of NMR experiments which purported to disprove this simple mechanism, but these have now been found to have been based on an incorrect assignment of NMR resonances.

Experience as a reviewer of grant applications has taught me that many scientists detest nothing more than simple explanations for complex phenomena. They would prefer cooperative oxygen binding to be due to transitions between a multitude of states of unknown structure rather than two structures capable of being determined by X-ray analysis; they would like the Bohr effect to arise from a host of small and inexplicable changes in pK_a: they would like the hydrogen bonds that you can see to make no signif-icant contribution to cooperativity, and those that you cannot see to be responsible for the cooperative effects.

By contrast, I believe that there is no longer any great mystery about the cooperative effects of vertebrate hemoglobins, and that the seemingly complex cooperative interactions in invertebrate hemoglobins and hemocyanins reported in this volume will find similarly simple explanations once their structure has been determined by X-ray analysis.

May, 1985   M.F. PERUTZ

Contents

List of Contributors

P. BILLIALD Laboratoire de Biochimie, Faculté de Pharmacie, Université François Rabelais, 37032 Tours Cedex, France

C. BONAVENTURA Marine Biomedical Center, Duke University Marine Laboratory, Beaufort, North Carolina 28516, U.S.A.

J. BONAVENTURA Marine Biomedical Center, Duke University Marine Laboratory, Beaufort, North Carolina 28516, U.S.A.

M. BRUNORI Istituto di Chimica, Facoltà di Medicina e Chirurgia, Università "La Sapienza", 00185 Roma, Italy

E. BURSAUX I.N.S.E.R.M. unité 27, 42 rue Desbassayns de Richemont, 92150 Suresnes, France

M. COLETTA Istituto di Chimica, Facoltà di Medicina e Chirurgia, Università "La Sapienza", 00185 Roma, Italy

F. GALACTEROS I.N.S.E.R.M., unité 91, Hôpital Henri Mondor, 51 Avenue du Maréchal de Lattre de Tassigny, 94010 CRETEIL, France

B. GIARDINA Istituto di Chimica, Facoltà di Medicina e Chirurgia, Università "La Sapienza", 00185 Roma, Italy

W.A. HENDRICKSON Department of Biochemistry and Molecular Biophysics, Columbia University, New York, NY 10032, U.S.A.

J. LAMY Laboratoire de Biochimie, Faculté de Pharmacie, Université François Rabelais, 37032 Tours Cedex, France

J. LAMY Laboratoire de Biochimie, Faculté de Médecine, Université François Rabelais, 37032 Tours Cedex, France

J. LEONIS Laboratoire de Chimie Générale I, Université Libre de Bruxelles, Av. F.D. Roosevelt, 50, B-1050 Bruxelles, Belgique

B. LINZEN Zoologisches Institut, Universität München, D-8000 München 2, R.F.A.

B. McMAHON Department of Biology, University of Calgary, Calgary, Alberta, Canada

K.I. MILLER Department of Biochemistry and Biophysics, Oregon State University, Corvallis, Oregon 97331, U.S.A.

G. MOTTA Laboratoire d'Immunogénétique, Université d'Orléans, 45046 Orléans, Cedex, and CNRS, C.S.E.A.L., 45045 Orléans, France

C. PAUL Laboratoire de Chimie Générale I, Université Libre de Bruxelles, Av. F.D. Roosevelt, 50, B-1050 Bruxelles, Belgique

D.A. POWERS Department of Biology and McCollum-Pratt Institute, The Johns Hopkins University, Baltimore, Maryland 21218, U.S.A.

C. POYART I.N.S.E.R.M. unité 27, 42 rue Desbassayns de Richemont, 92150 Suresnes, France

W. SCHARTAU Zoologisches Institut, Universität München, D-8000 München 2, R.F.A.

H.-J. SCHNEIDER Zoologisches Institut, Universität München, D-8000 München 2, R.F.A.

A.G. SCHNEK Laboratoire de Chimie Générale I, Université Libre de Bruxelles, Av. F.D. Roosevelt, 50, B-1050 Bruxelles, Belgique

S. SHERIFF Genex Corporation, Gaithersburg, Maryland 20877, and Laboratory for the Structure of Matter, Naval Research Laboratory , Washington, D.C. 20375, U.S.A.

P.-Y. SIZARET Laboratoire de Microscopie Electronique, Faculté de Médecine, Université François Rabelais, 37032 Tours Cedex, France

J.L. SMITH Laboratory for the Structure of Matter, Naval Research Laboratory, Washington, D.C. 20375, U.S.A.

K.E. VAN HOLDE Department of Biochemistry and Biophysics, Oregon State University, Corvallis, Oregon 97331, U.S.A.

S.N. VINOGRADOV Biochemistry Department, Wayne State University School of Medicine, Detroit, Michigan 48201, U.S.A.

H. WACJMAN I.N.S.E.R.M. unité 15, Institut de Pathologie Moléculaire, C.H.U. Cochin Port-Royal, Rue du Faubourg St Jacques, 75014 Paris, France

L. ZOLLA Istituto di Chimica, Facoltà di Medicina e Chirurgia, Università "La Sapienza", 00185 Roma, Italy

Structure and Function of Hemerythrins

W.A. HENDRICKSON, J.L. SMITH, S. SHERIFF

I. INTRODUCTION

Nature has evolved three fundamentally different molecular devices to serve as oxygen carriers in support of animal respiration. The striking color and abundant supply of these proteins has made them frequent subjects of study. Hemoglobin is the familiar red substance in the blood of humans and many other animals; hemocyanin is the blue pigment in the blood of many molluscs and arthropods; and hemerythrin is the burgundy colored protein in the body fluids of a few minor invertebrate phyla. There is considerable diversity in the physiological parameters and in the structure and symmetry of hemoglobins and hemocyanins. It appears from recent work that the less thoroughly studied hemerythrin family also exhibits such diversity although the basic framework of the protein and active center appear to be conserved through evolution.

In this article we give a brief review of the properties of hemerythrins and summarize some of the findings from our structural studies of these molecules by X-ray crystallography and other physical techniques.

II. CHEMISTRY AND FUNCTION

Hemerythrins have been found in sipunculans and priapulids, in certain brachiopods and in at least one genus of annelids. In all cases the spectroscopic properties and chemical reactions of these proteins are quite similar. These characteristics have been the subject of several reviews, most recently by Klotz and Kurtz (1984), and are the consequence of the chemistry at a dimeric iron active center. Deoxyhemerythrin is colorless and has Mössbauer parameters characteristic of high-spin Fe(II) whereas upon reaction with oxygen the protein takes on a burgundy color and the Mössbauer and magnetic susceptibility data indicate antiferromagnetically coupled high-spin Fe(III) centers. The $Fe:O_2$ stoichiometry is 2:1. Upon prolonged standing oxyhemerythrin autoxidizes to a met state that is yellowish green under acidic conditions (pH < 7) and more brownish yellow when alkaline (pH > 8). Met-hemerythrin reacts with various anionic ligands, such as Cl^-, CN^-, and N_3^-, and these ions

2

also promote a rapid conversion of oxyhemerythrin to the met state. The colors of these complexes are variable depending on charge transfer bands. Methemerythrin can also be produced by oxidation with ferricyanide, and methemerythrin can be reduced to the deoxy state with thiocyanate. Under controlled conditions these reactions can be stopped midway to yield semi-met hemerythrin with mixed Fe(II)-Fe(III) iron centers (Babcock et al., 1980).

The protein exerts substantial control on the reactivity of the hemerythrin active center. Such modulation is essential to the physiological needs of the animals. The oxygen affinity must be so modulated that effective transfer of oxygen from the supply in bathing sea water to the locus of need at the mitochondria of respiring tissues occurs. In the sipuncula, this transfer is subserved by three types of hemerythrin -- vascular, coelomic, and muscular. At least in the genus Themiste, vascular blood is pumped through the tentacles in contact with the sea and on through vessels to the coelom. The coelomic fluid and its erythrocytes in turn bathe the retractor muscles that contain a myohemerythrin (Manwell, 1960). Klippenstein et al. (1972) showed that Themiste zostericola hemerythrins from these three compartments are in fact distinct proteins. They also have distinctive functional properties as shown in figure 1. The oxygen affinities are compatible with a system of transfer from the vascular system into the coelomic fluid and on into the muscle tissues. Other functional relations exist in other species which live in different environments (Manwell, 1960). Isolated hemerythrins show little cooperativity in oxygen binding (Hill coefficients under 1.3), but there is evidence of large cooperativity in vivo (Mangum and Kondon, 1975) suggesting modulation by effector agents.

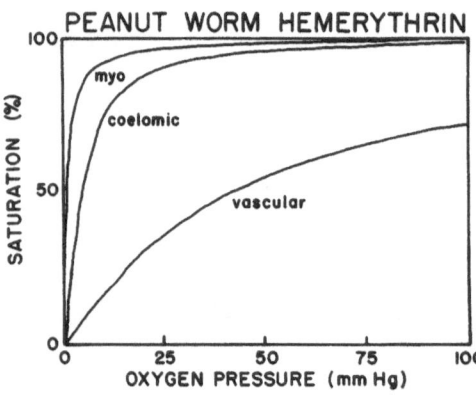

FIG. 1. Oxygen binding equilibria for The-miste zostericola hemerythrins. Parameters used in drawing these curves are based on measurements reported by Manwell (1960) for the vascular and coelomic proteins and measurements on myohemerythrin by GL Klippenstein (personal communication, 1979). The values used are as follow for n and P_{50} respectively: 1.1 and 42.0 for vascular hemerythrin, 1.3 and 4.5 for coelomic hemerythrin, and 1.0 and 0.9 for myohemerythrin.

III. PROTEIN STRUCTURE AND ASSOCIATION

All available evidence is consistent with a universal "hemerythrin fold". Molecular weights and dissociation properties of all well characterized hemerythrins are in multiples 13-14,000 dalton units. Also, the amino acid sequences that have been determined all have either 113 or 118 amino acid residues. Finally, four X-ray crystal structures have been shown to have the same tertiary structure. The basic framework of this structure is a bundle of four essentially antiparallel alpha-helices. In addition there is an N-terminal tail in an extended conformation and a C-terminal stub in a tight conformation that closes one end of the bundle. The iron center is located about one-third of the way along the helices from the C-terminal end.

The polypeptide conformation and the known modes of association are shown in figure 2. These crystal structures are of the monomeric myohemerythrin from Themiste zostericola (Hendrickson et al., 1975), of the octameric coelomic hemerythrin from Phascolopsis gouldii (Ward et al., 1975), and of the trimeric hemerythrin from Siphonosoma funafuti or a closely related species (Smith et al., 1983). The crystal structure of another octameric hemerythrin has also been determined (Stenkamp et al., 1976). All of these hemerythrins are from Sipunculan worms. It has been thought earlier that hemerythrins were generally octameric. However, as more animals are studied increasing diversity appears to exist. The first trimeric hemerythrin to be characterized was found in Phascolosoma agassizii (Liberatore et al., 1974). Since then the trimeric nature of S. funafuti hemerythrin has been established and hemerythrin from S. cumanense exists as trimers and higher aggregates (A.W. Addison, personal communication). Crystals of this hemerythrin are consistent with the existence of hexameric hemerythrin. However, there is no clear phylogenetic pattern in the degree of association. Whereas the hemerythrin from P. agassizii is trimeric, that from P. arcuatum appears to be a mixture of dimers and tetramers (Manwell, 1977) and the dimeric form has been crystallized (Sieker et al., 1981).

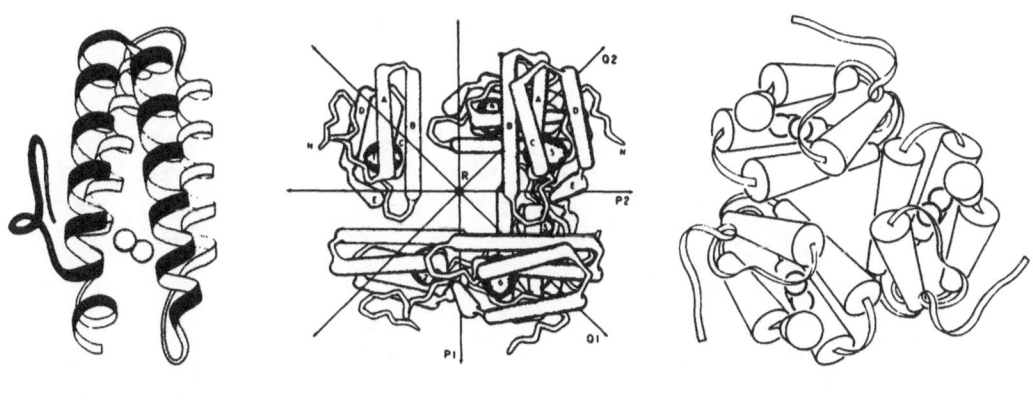

MYOHEMERYTHRIN　　　**OCTAMER**　　　　**TRIMER**

FIG. 2. Schematic representations of three hemerythrin structures. The myohemerythrin drawing is by JS Richardson, the octamer drawing is by D Ward and the trimer drawing is by JC Moon.

Diversity in association also extends to other phyla. The coelomic hemerythrins from the brachiopod Lingula unguis appear to be octameric (Joshi and Sullivan, 1973). On the other hand, those from the priapulid Priapulus caudatus appear to be based on the trimeric pattern. We have recently purified this hemerythrin and characterized it by Sephadex gel filtration and SDS polyacrylamide gel electrophoresis (Smith and Hendrickson, unpublished results). The main component of the material appears to be a trimer (37,000 daltons) in equilibrium with the monomer (14-15,000 daltons). At high concentration, hexamers seem to form. A non-aggregating hemerythrin component of low molecular weight (≈10,000 daltons) is also present.

IV. ACTIVE CENTER STRUCTURE AND MECHANISM

A detailed understanding of the process of reversible oxygenation in hemerythrin requires a knowledge of the structure of the active center. We have recently refined the structure of myohemerythrin at high resolution (1.3 Å in one direction and 1.7 Å in the other two). This structure is in the azidomet state. Stenkamp et al. (1981, 1982) have also refined the atomic model for the octameric hemerythrin from T. dyscrita in the azidomet and acid met states at 2.0 Å resolution. These results give a definitive picture of the center and a stereodrawing is presented in figure 3. The two azidomet structures are virtually identical in the active center.

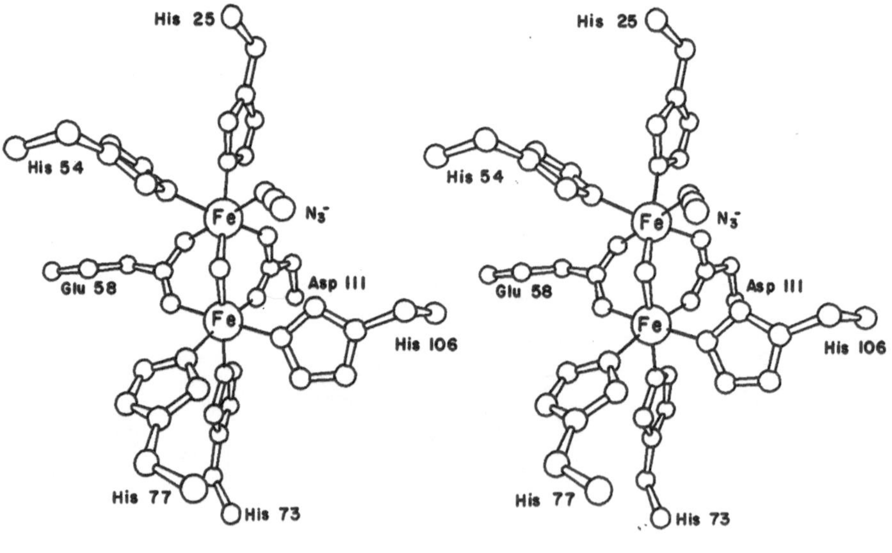

FIG. 3. Stereodrawing of the active center in myohemerythrin.

Each iron atom is coordinated octahedrally by oxygen and nitrogen atoms. An oxide ion with the short bonds (≈1.8 Å) of a μ-oxo bridge is responsible for the antiferromagnetic coupling between the iron atoms. One iron atom is further coordinated by three histidines (residues 73,77 and 106) and the other has two histidine ligands (residues 25 and 54) and the exogenous azide ligand.

X-ray absorption spectroscopy, and particularly extended X-ray absorption (EXAFS) analyses are consistent with the above picture of the hemerythrin iron center (Elam et al., 1982; Hendrickson et al., 1982). In addition, the EXAFS investigations provide structural information about the physiologically relevant oxy and deoxy states that have not been thoroughly studied by crystallography. We find that the μ-oxo bridge that is present in azidomethemerythrin also persists in oxyhemerythrin, but that the short (1.8 Å) oxo bridge is not present in deoxyhemerythrin (Smith, Co, Hendrickson and Hodgson, unpublished results).

The detailed atomic model of the azidomet active center together with data on the oxy and deoxy states from X-ray absorption spectroscopy has provided the basis for constructing a rather detailed reaction mechanism. Some features of this mechanism are shown in figure 4. In addition to the specific structural data, three other crucial facts were used: (1) for many hemerythrins there is no Bohr effect, i.e. oxygenation proceeds independent of pH (Bates et al., 1968), (2) oxygen binds as the peroxide ion, O_2^{2-} (Dunn et al., 1975) and (3) the μ-oxo bridge oxygen does not exchange with water during oxygenation or deoxygenation (Frier et al., 1980). These data are consistent with a deoxy hemerythrin structure that includes a hydroxyl bridge as shown and an attack by oxygen that involves electron withdrawal from the metal centers and internal proton transfer from the hydroxyl to the oxygen molecule to produce hydroperoxide. The hydroperoxide complex (oxyhemerythrin) is stabilized by a hydrogen bond to the μ-oxo bridge. This hydrogen bond provides an explanation for the differences between the resonance Raman spectra, magnetic susceptibility data and Mössbauer spectra of oxyhemerythrin in comparison to those from methemerythrin. The proposed oxy and deoxy structures have feasible geometry when built into the azidomet framework.

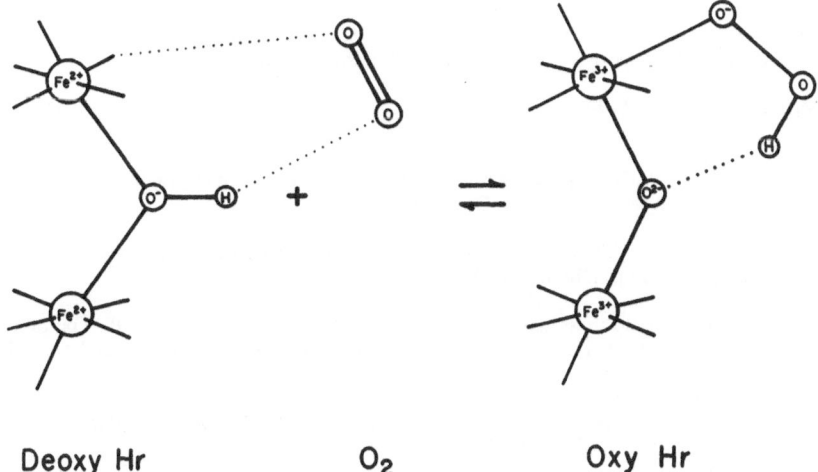

Deoxy Hr O_2 Oxy Hr

FIG. 4. Aspects of a proposed reaction mechanism for reversible oxygenation of hemerythrin.

The proposals of this mechanism require tests such as ones from X-ray crystallography and nuclear magnetic resonance that are underway. A detailed understanding of the mechanism will facilitate the investigation of the origin of the biochemical adaptation that occurs in response to differing physiological demands.

REFERENCES

Babcock LM, Brandic Z, Harrington PC, Wilkins RG, Yoneda GS (1980) Preparation, disproportionation, and reactions of two semi-met forms of hemerythrin. J. Amer. Chem. Soc. 102: 2849-2850

Bates G, Brunori M, Amiconi G, Antonini E, Wyman J (1968) Studies on hemerythrin. I. Thermodynamic and kinetic aspects of oxygen binding. Biochemistry 7: 3016-3020

Dunn JBR, Shriver DF, Klotz IM (1975) Resonance Raman studies of hemerythrin-ligand complexes. Biochemistry 14: 2689-2695

Elam WT, Stern EA, McCallum JD, Sanders-Loehr J (1982) Structure of the binuclear iron center in hemerythrin by X-ray absorption spectroscopy. J. Amer. Chem. Soc. 104: 6369-6373

Freier SM, Duff LL, Shriver DF, Klotz IM (1980) Resonance Raman spectroscopy of iron-oxygen vibrations in hemerythrin. Arch. Biochem. Biophys. 205: 449-463

Hendrickson WA, Klippenstein GL, Ward KB (1975) Tertiary structure of myohemerythrin at low resolution. Proc. Natl. Acad. Sci. USA 72: 2160-2164

Hendrickson WA, Co MS, Smith JL, Hodgson KO, Klippenstein GL (1982) X-ray absorption spectroscopy of the dimeric iron site in azidomethemerythrin from Phascolopsis gouldii. Proc. Natl. Acad. Sci. USA 79: 6255-6259

Joshi JG, Sullivan B (1973) Isolation and preliminary characterization of hemerythrin from Lingula urguis. Comp. Biochem. Physiol. 44B: 857-867

Klippenstein GL, Van Riper DA, Oosterom EA (1972) A comparative study of the oxygen transport proteins of Dendrostomurn pyroides. J. Biol. Chem. 247: 5959-5963

Klotz IM, Kurtz DM, Jr. (1984) Binuclear oxygen carriers: hemerythrin. Account Chem. Res. 17: 16-22

Liberatore FA, Truby MF, Klippenstein GL (1974) The quaternary structure of Phascolopsis agassizii coelomic hemerythrin. Arch. Biochem. Biophys. 160: 223-229

Mangum CP, Kondon K (1975) The role of coelomic hemerythrin in the sipunculid worm Phascolopsis gouldii. Comp. Biochem. Physiol. 50A: 777-785

Manwell C (1960) Histological specificity of respiratory pigments. - II. Oxygen transfer systems involving hemerythrins in sipunculid worms of different ecologies. Comp. Biochem. Physiol. 1: 277-285

Manwell C (1977) Superoxide dismutase and NADH diaphorase in hemerythrocytes of sipunculans. Comp. Biochem. Physiol. 56B: 331-338

Sieker LC, Bolles L, Stenkamp RE, Jensen LH, Appleby CA (1981) Preliminary X-ray study of a dimeric form of hemerythrin from Phascolosoma arcuatum. J. Mol. Biol. 148: 493-494

Smith JL, Hendrickson WA, Addison AW (1983) Structure of trimeric hemerythrin. Nature 303: 86-88

Stenkamp RE, Sieker LC, Jensen LH, Loehr JS (1976) Structure of methemerythrin at 5 Å resolution. J. Mol. Biol. 100: 23-34.

Stenkamp RE, Sieker LC, Jensen LH, Sanders-Loehr J (1981) Structure of the binuclear iron complex in metazido hemerythrin from Themista dyscritum at 2 A resolution. Nature 291: 263-264

Stenkamp RE, Sieker LC, Jensen LH (1982) Restrained least-squares refinement of Themiste dyscritum methydroxohemerythrin at 2.0 A resolution. Acta Cryst. B38: 784-792

Ward KB, Hendrickson WA, Klippenstein GL (1975) Quaternary and tertiary structure of hemerythrin. Nature: 257: 818-821

The Structure of Erythrocruorins and Chlorocruorins, the Invertebrate Extracellular Hemoglobins

S.N. VINOGRADOV

I. INTRODUCTION

The intermittent occurrence of intracellular and extracellular hemoglobins among the invertebrates may be due either to the intermittent presence of functional globin genes or to defects in the mechanism regulating their expression. In the absence of information concerning the structure and expression of invertebrate globin genes, I would like to focus on the identification of common structural themes among these molecules. All the extracellular invertebrate hemoglobins whose subunit structures are known can be classified into four separate groups: (A) single-domain, single-subunit molecules, i.e. monomeric hemoglobins, (B) two-domain, multi-subunit hemoglobins, consisting of aggregates of dimeric polypeptide chains, each containing two heme-binding domains, (C) multi-domain, multi-subunit hemoglobins, consisting of two or more polypeptide chains, each comprising many heme-binding domains and (D) single domain, multi-subunit hemoglobins, consisting of aggregates of monomeric and dimeric subunits and disulfide-bonded trimers or tetramers. A diagrammatic representation of the subunits found in the four classes is shown in figure 1.

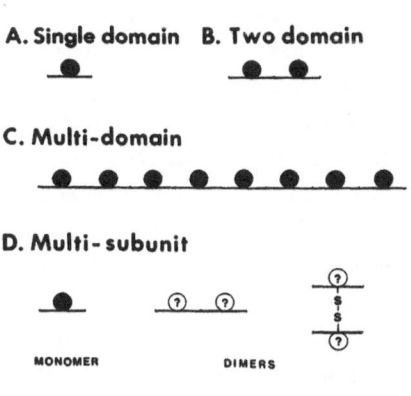

FIG. 1. Diagrammatic representation of the subunits of the four classes of invertebrate extracellular hemoglobins: (A) single-domain, single-subunit, (B) two-domain, multi-subunit, (C) multi-domain, multi-subunit, and (D) single-domain, multi-subunit. The filled circles represent the heme-binding domains.

II. SINGLE-DOMAIN, SINGLE SUBUNIT HEMOGLOBINS

Hemoglobin is known to exist in ciliated protozoa such as Paramecium and Tetrahymena (Keilin and Ryley, 1953). Paramecium hemoglobin has been shown to be monomeric and polymorphic (Steers and Davis, 1979; Irie and Usuki, 1980). Hemoglobin occurs in many tissues in a variety of parasitic and free-living platyhelminths and nematodes (Lee and Smith, 1965; Ellenby and Smith, 1966; Phillips, 1978). The hemoglobin of the parasitic platyhelminth Dicrocoelium dendriticum is known to be monomeric (Tuchsmid et al., 1978). Among the insects, hemoglobin is very rare: it occurs in a few Hemiptera such as Buenoa confusa (Bergstrom, 1977) and among the Diptera, only in the larva of the midge Chironomus (Braun et al., 1968).

Hemoglobin occurs in the nerve tissue of many invertebrates scattered among the nemerteans, annelids, arthropods and molluscs. The nerve tissue hemoglobins of the polychaete Aphrodite aculeata and of the gastropod mollusc Aplysia californica have been shown to be monomeric (Wittenberg et al., 1965).

The most studied monomeric hemoglobins are undoubtedly the polymorphic hemoglobins of Chironomus (Braun et al., 1968). The amino acid sequences of all twelve hemoglobins of Chironomus thummi thummi have been determined by G. Braunitzers' group (Goodman et al., 1983). Seven of these hemoglobins have a capacity for dimerization and the genes for the two types of globin chains are clustered in two different chromosomal regions.

III. TWO-DOMAIN, MULTI-SUBUNIT HEMOGLOBINS

The smallest subunit of these molecules is a polypeptide chain of 35 ± 5 kDa which contains two separate, linearly connected, heme-binding domains (see figure 1B). Such a chain is probably the result of the duplication of an ancestral globin gene. The two-domain subunits aggregate into large molecules with molecular masses ranging from 250 kDa to 800 kDa. These hemoglobins are found among the carapaced branchiopod crustaceans (Ilan and Daniel, 1979a; Daniel, 1983). The hemoglobins of Cyzicus hierosolymitanus and of Caenestheria inopinata, from the order Conchostraca, have a molecular mass of ca. 300 kDa and contain 10 chains (Ilan et al., 1981); the hemoglobin of Daphnia magna, from the order of Cladocera, has a mass of 490 kDa and contains 16 chains (Ilan and Daniel, 1982) and that of Lepidurus apus, from the order Notostraca, has a mass of 800 kDa and contains 24 chains (Ilan and Daniel, 1979b). In electron micrographs Cyzicus hemoglobin appears to consist of two stacked pentagonal rings, 13 ± 1 nm in diameter and 8.5 ± 0.5 nm high (David et al., 1977). Although Caenestheria hemoglobin also has a pentagonal shape, Daphnia hemoglobin appears as a two-tiered ring, ca. 14 nm in diameter, each tier consisting of eight subunits.

A dimeric hemoglobin occurs in the larva of the hemipteran Gastrophilus intestinalis; it has a molecular mass of 35 kDa and a minimum molecular mass based on heme content, of

17.6 kDa (Phelps et al., 1972). The hemoglobin of the nematode Ascaris lumbricoides possesses a mass of 330 kDa and consists of subunits of 38-40 kDa which, however, bind only one heme group instead of two (Okazaki et al., 1965). Possibly alterations in the amino acid sequence of one of the two heme-binding domains has resulted in the loss of the ability to bind heme. The polymeric hemoglobin which exists together with a tetrameric hemoglobin, in the erythrocytes of the arcid clam Barbatia reevana is another hemoglobin which belongs to this class: its mass of ca. 430 kDa makes it the largest known intracellular hemoglobin (Grinich and Terwilliger, 1980).

IV. MULTI-DOMAIN, MULTI-SUBUNIT HEMOGLOBINS

This group consists of two or more polypeptide chains each containing from 8 to 20 heme-binding domains (see figure 1C). Such chains must have arisen as the result of repeated duplications of an ancestral globin gene. These hemoglobins are found in carapace-less branchiopod crustaceans of the order Anostraca, in planorbid snails and in clams of the families Astartidae and Carditidae.

Artemia salina contains three hemoglobins which have a mass of ca. 250 kDa and consist of two polypeptide chains, each containing 8 heme-binding domains (Moens and Kondo, 1978; Wood et al., 1981). Hb I and Hb III are homodimers of two chemically distinct chains α and β, respectively, and Hb II is an $\alpha \beta$ heterodimer (Moens, 1982). In electron micrograph, Artemia hemoglobin appears as a circular, lobed structure, 12 ± 1 nm in diameter and 7 ± 1 nm in height.

The hemoglobins of the snails Planorbis corneus and Helisoma trivolvis have masses of about 1750 kDa and both appear as circular arrays in electron micrographs (Wood and Mosby, 1975; Terwilliger et al., 1976a). Although Planorbis hemoglobin exhibits a ring-like structure 12.2 nm in diameter possessing hexagonal symmetry, Helisoma hemoglobin has an overall diameter of ca. 20 nm and appears to possess tenfold symmetry in agreement with the fact that its polypeptide chain has a mass of about 175 kDa. The minimum molecular mass based on heme content is ca. 18 kDa for Helisoma hemoglobin and 22.3 kDa for Planorbis hemoglobin.

The extracellular hemoglobins of the clams Cardita affinis, C. borealis and Astarte castanea appear in electron micrographs as rod-like polymers with diameters ranging from 21 to 37 nm and lengths varying from 37 to 120 nm (Terwilliger et al., 1978; Terwilliger and Terwilliger, 1978; Yager et al., 1982). These aggregates are formed of ca. 300 kDa polypetide chains containing up to 20 heme-binding domains (Terwilliger and Terwilliger, 1983). The minimum molecular masses of the clam hemoglobins based on heme content, are 17 to 20 kDa.

A characteristic feature of this class is that when the molecule is digested briefly with a nonspecific protease such as subtilisin, it yields heme-containing fragments that are integral multiples of 15-17 kDa and which retain the capacity for binding oxygen.

V. SINGLE-DOMAIN MULTI-SUBUNIT HEMOGLOBINS

These molecules possess a characteristic two-tiered hexagonal appearance and are ubiquitous among all three classes of annelids. They also comprise the chlorocruorins, found in four polychaete families, whose heme group possesses a formyl instead of a vinyl group in position 2. All the molecules possess a sedimentation coefficient in the range of 52 to 61S, an acidic isoelectric point and a low iron content of 0.24 ± 0.03% corresponding to a minimum molecular mass of 23 ± 3 kDa. The physicochemical properties of the few chlorocruorins that have been studied appear to be very similar to the corresponding properties of the hemoglobins (Terwilliger et al., 1976b). The similarity between annelid hemoglobin and chlorocruorin is underscored by the polychaete Serpula vermicularis which contains both in the ratio of 3 to 2; the two molecules cannot be separated and the minimum molecular mass based on the iron content is 24.7 kDa (Terwilliger, 1978).

From the early studies by Svedberg to the present, the published molecular masses of the annelid hemoglobins have ranged from 2,200 to 3,800 kDa (Chung and Ellerton, 1979; Vinogradov et al., 1980). Surprisingly, the range of the reported values for the molecular mass of earthworm hemoglobin is equally broad (Vinogradov et al., 1977). Recently, sedimentation equilibrium values and the values obtained by small angle X-ray scattering in solution have tended to cluster in the upper portion of the range, between 3,500 to 4,000 kDa (Pilz et al., 1980; Wood et al., 1976). At present, it is not possible to say whether there are any differences in molecular mass between various annelid hemoglobins.

Conventional transmission electron microscopic (CTEM) studies of annelid hemoglobins have tended to suggest that their vertex-to-vertex diameters were 24 to 27 nm and their heights were 15 to 18 nm (Vinogradov et al., 1982). Scanning transmission electron microscope (STEM) studies of several annelid molecules have revealed some differences. Although the dimensions of the hemoglobins of Lumbricus terrestris, Arenicola marina, Nephtys incisa, Tubifex tubifex and Macrobdella decora are all 30 x 20 nm (Kapp et al., 1982; Kapp and Crewe, 1984; Messerschmidt et al., 1983), those of Myxicola infundibulum chlorocruorin were 30 x 8 nm (Vinogradov et al., 1984) and those of Amphitrite ornata hemoglobin were 26 x 17 nm (Kapp and Crewe, personal communication). Some photographs of the digitized STEM images are shown in figure 2. We do not know whether the differences in molecular size obtained by STEM correlate with differences in molecular mass. The vertex-to-vertex dimension of 30 nm obtained by STEM is in agreement with the hexagonal spacing obtained recently for wet crystals of Lumbricus hemoglobin using X-ray diffraction (Royer et al., 1981).

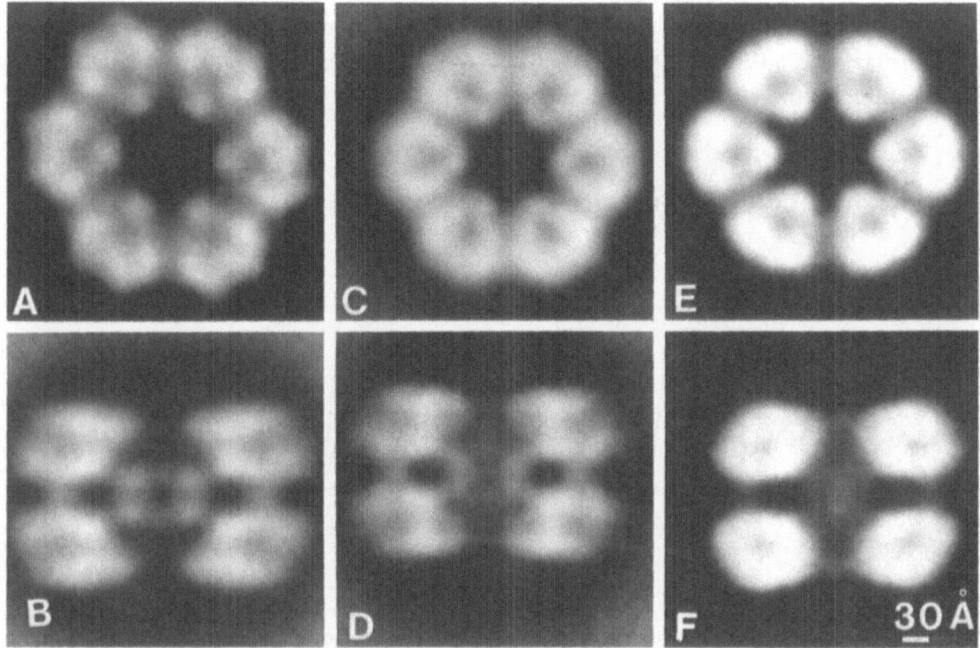

FIG. 2. Summed and averaged digitized STEM images of the top and side-views of
Lumbricus terrestris hemoglobin (A, B), Myxicola infundibulum chlorocruorin (C, D) and
Amphitrite ornata hemoglobin (E, F).

VI. SUBUNIT STRUCTURE OF ANNELID HEMOGLOBINS AND CHLOROCRUORINS

All the oligochaete and polychaete hemoglobins examined so far exhibit a uniform
unreduced SDS PAGE pattern: a monomer comprising 20% of the total, a trimer comprising
60% of the total and several dimeric subunits of ca. 33 to 37 kDa resistant to reduction,
accounting for the remainder (Vinogradov et al., 1976, 1980). The only exception is the
hemoglobin of Nephtys incisa which appears to consist predominantly of disulfide-bonded 50
kDa trimers and 65 kDa tetramers (Vinogradov et al., 1983). The unreduced SDS PAGE
patterns of the achaete hemoglobins consists of monomeric and tow types of dimeric
subunits: a disulfide-bonded dimer and chains of 30 to 37 kDa which did not dissociate upon
reduction (Andonian and Vinogradov, 1975; Andonian et al., 1975; Shlom et al., 1975). The
unreduced chlorocruorins exhibit two different patterns: one consisting of monomers and
disulfide-bonded dimers and tetramers as in the case of Potamilla leptochaeta (Vinogradov
and Orii, 1980) and Spirographis (Di Stefano et al., 1977; Mezzasalma et al., 1983)
chlorocruorins and the other consisting of disulfide-bonded tetramers comprising at least
80% of the total and a chain of ca. 30 kDa unaffected by reduction, as in the case of

Myxicola chlorocruorin (Vinogradov et al., 1985). Figure 3 shows a diagram of the representative SDS PAGE patterns. The diversity of the subunits shown in figure 1D suggests that at present, there may be several branches in the evolution of this group of molecules.

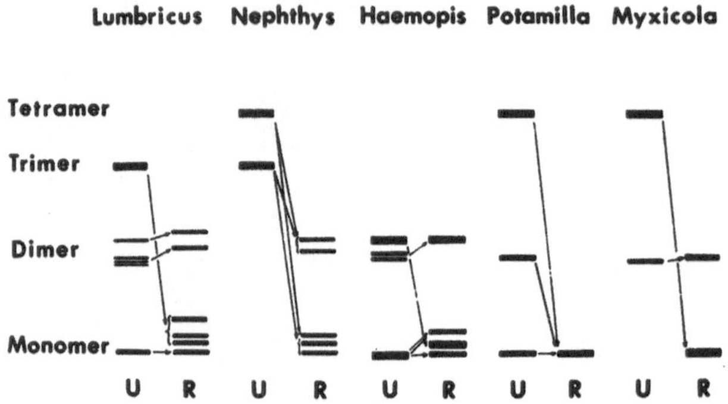

FIG. 3. Diagrammatic representation of the SDS PAGE patterns of unreduced (U) and reduced (R) Lumbricus terrestris, Nephthys incisa and Hemopis grandis hemoglobins and of Potamilla leptochaeta and Myxicola infundibulum chlorocruorins. The arrows indicate the relationships between the two sets of subunits found by reelectrophoresis or two-dimensional SDS PAGE.

The annelid hemoglobins and chlorocruorins differ from the other extracellular hemoglobins in possessing a number of polypeptide chains greater than the number of heme groups. Lumbricus hemoglobin has a molecular mass of about 4,000 kDa, contains 156 ± 5 heme groups and consists of at least six different subunits: I through IV possessing a mass of 16-19 kDa and subunits V and VI having a mass of 33 and 37 kDa, respectively (Shlom and Vinogradov, 1973); the total number of polypeptide chains comprising the hemoglobin must be close to 200. A similar estimate was arrived at in the case of hemoglobin of a polychaete, Tylorrhynchus heterochaetus (Gotoh and Kamada, 1980).

VII. RELATEDNESS OF INVERTEBRATE AND VERTEBRATE HEMOGLOBINS

The amino acid sequences of the small monomeric subunit of Tylorrhynchus hemoglo-bin (Suzuki et al., 1982) and of one of the chains of Lumbricus hemoglobin (Garlick and Riggs, 1982) have been determined. In Lumbricus hemoglobin the heme group is inserted in the same orientation as in mammalian hemoglobin judging from the ratio of biliverdin IX α and β isomers formed upon coupled oxidation (Brown, private communication; Docherty and Brown, 1982).

Figure 4 compares the two extracellular annelid sequences with the primary structures of the intracellular, monomeric hemoglobin of the polychaete Glycera dibranchiata (Imamu-

```
No.AA Mol.Mass

                                                    A                                   B         C
LUMBRICUS      157  17,496   K K Q C G V L E G L K V K S E W G R A - - - Y G S G H D R E A F S Q A I W R A T F
TYLORRHYNCHUS  139  16,327   T D C G I L Q R I K V K Q Q W A Q V - - - Y S V G E S R T D F A I D V F N N F F
GLYCERA        147  15,590   G - L S A A Q R Q V I A A T W K D I A G N D N G A G V G K D C L I K H L S A H P Q
C T T III      136  15,400   L S A D Q I S T V Q A S F D K V - - K G D P V G - - - I L Y A V E K A D P S
BETA CHAIN     146  15,867   V H L T P E E K S A V T A L W G K V - - N V D E V G - - G E A L G R L L V V Y P W
                             C CD           CD D           DE                  EF F'        F' F

                                                    A                                   B
LUMBRICUS      A Q V P E S R S L F K R V H G D H T S D P A F I A H A E R V L G G L D I A I S T L D Q P A T L K E E L
TYLORRHYNCHUS  R T N P D - R S L F N R V N G D N V Y S P E F K A H M V R V F A G F D I L S V L D D D K P V L D Q A L
GLYCERA        M A A V F G - F S G - A S - - I - - D P A V A D L G A K V L A Z I G V A V S H L G D Z G K M V A Q M
C T T III      I M A K F T Q F A G K D L E S I K G T A P F E - T H A N R I V G F F S K I G E L P N I E A D - - V
BETA CHAIN     T Q R F F E S F G D L S T P D A V M G N P K V K A H G K K V L G A F S D G L A H L D N - - L K G T F
               F FG            FG G              G GH                 GH                    H

                                                    A                                   B
LUMBRICUS      D H L Q V Q H E G R - - K - I P D - N Y F D A F K T A I L H V V A A Q L G E R G Y S N N E E I H D A I
TYLORRHYNCHUS  A H Y A A F H K Q F - - G T I P F - K A F G Q T - - - M F Q T I A E H I - - - - - - H G A
GLYCERA        K A V G V B H K G Y G N K - H I K G Q Y F E P L G A S L L S A M E H R I G G K M - - - - N A A
C T T III      N T F V A S H K P R G V T H D Q - L N N F R A G F V S Y M K A H T D F A G A E - - - - - -
BETA CHAIN     A T L S E L H C D K - - L - H V D P E N F R L L G N V L V C V L A H H F G K E F - - - - T P P

                                                                 H
LUMBRICUS      A C D G F A R V L P Q V L E R G I K G H H
TYLORRHYNCHUS  D I G A W R A C Y A E Q I V T G I T A
GLYCERA        A K D A W A A A Y A D I S G A L I S G L Q S
C T T III      - - A A W - G A T L D T F F G M I F S K M
BETA CHAIN     V Q A A Y K V V A Q V A N A L A H K Y
```

FIG. 4. Amino acid sequences of a globin chain of Lumbricus terrestris hemoglobin, the monomeric globin chain of Tylorrhunchus heterochaetus, the intracellular monomeric globin of Glycera dibranchiata, component III of Chironomus thummi thummi hemoglobin and the human β chain.

ra et al., 1972), of component III of Chironomus hemoglobin (Buse et al., 1979) and of the human β chain. The invertebrate globin sequences are very distant from the human α and β chains. Surprisingly, although the number of amino acid identities between the Lumbricus sequence and the sequences of the human α and β chains and of the Glycera chain are 20, 25 and 24, respectively, there are only 15 identities with the CTT III sequence. The Tylorrhynchus sequence has 25 amino acid identities with the Glycera sequence, 42 identities with the Lumbricus sequence and only 18 with the CTT III sequence. It is noteworthy that the first 105 residues of the Tylorrhynchus sequence exhibit almost all of the amino acid identities with the Lumbricus sequence, while the majority of the identities with the Glycera sequence occur in the C-terminal portion. Obviously much more data is needed on the sequences of invertebrate globin chains before any reliable conclusions can be drawn.

X-ray crystal structure analyses of CTT III (Steigemann and Weber, 1979) of the Glycera hemoglobin (Padlan and Love, 1974) and a leghemoglobin (Vainshtein, 1981) have demonstrated the identity of the secondary and tertiary structures of these monomeric globins with those of the vertebrate chains despite the substantial differences in primary structures. It will be interesting to see the three-dimensional folding of the two-domain and the multi-domain globin chains as well as of the unusal dimers and disulfide-bonded trimers and tetramers observed in the annelid hemoglobins and chlorocruorins.

VIII. CONCLUSION

When the subunit structures of the invertebrate extracellular hemoglobins are compared with those of the extracellular hemocyanins of the invertebrates, there emerges a considerable similarity between the two. The two dominant structural themes found among the extracellular hemoglobins, namely, the aggregation of multi-domain polypeptide chains on one hand and the aggregation of single-domain subunits, on the other, are also found among the hemocyanins. The molluscan hemocyanins consist of chains of about 400 kDa, each containing 8 copper-binding sites, which aggregate into large cylindrical structures. The arthropod hemocyanins consist of variable aggregates of polypeptide chains of about 90 kDa, each containing one copper-binding site (Ellerton et al., 1983; van Holde and Miller, 1982).

ACKNOWLEDGMENT

Some of the work described in this review was supported in part by United States Public Health Services, Grant HL 25952.

REFERENCES

Andonian MR, Vinogradov SN (1975) Physical properties and subunits of Dina dubia erythrocruorin. Biochim. Biophys. Acta 400: 344-354

Andonian MR, Barrett AS, Vinogradov SN (1975) Physical properties and subunits of Haemopis grandis erythrocruorin. Biochim. Biophys. Acta 412: 202-213

Bergtrom G (1977) Partial characterization of hemoglobin of the bug Buenoa confusa. Insect Biochem. 7: 313-316

Braun V, Crichton R, Braunitzer G (1968) Hämoglobine XV. Uber monomere und dimere Insekthämoglobine. Hoppe-Seyler's Z. Physiol. Chem. 349: 197-210

Buse G, Stettens GJ, Braunitzer G, Steer W (1979) Hämoglobine XXV. Hämoglobin (Erythrocruorin) CTT III aus Chironomus thummi thummi; primarstruktur und Beziehung zu anderer Hemproteine. Hoppe-Seyler's Z. Physiol. Chem. 360: 89-97

Chung MCM, Ellerton HD (1979) The physicochemical and functional properties of extra-cellular respiratory haemoglobins and chlorocruorins. Progr. Biophys. Mol. Biol. 35: 53-102

Daniel E (1983) Subunit structure of arthropod erythrocruorins. Life Chemistry Reports, suppl. 1: 157-166

David MM, Schejter A, Daniel E, Ar A, Ben-Shaul Y (1977) Subunit structure of hemoglobin from the clam shrimp Cyzicus. J. Mol. Biol. 111: 211-214

Di Stefano L; Mezzasalma V, Piazzese S, Russo GC, Salvato B (1977) The subunit structure of chlorocruorin. FEBS Lett. 79: 337-340

Docherty JC, Brown SB (1982) Haem disorder in reconstituted human hemoglobin. Biochem. J. 207: 583-587

Ellenby C, Smith L (1966) Haemoglobin in Mermis subnigrescens, Enoplus brevis and E. communis. Comp. Biochem. Physiol. 19: 871-877

Ellerton HD, Ellerton NF, Robinson HA (1983) Haemocyanin - a current perspective. Progr. Biophys. Mol. Biol. 41: 143-248

Garlick RL, Riggs A (1982) The amino acid sequence of a major polypeptide chain of earthworm hemoglobin. J. Biol. Chem. 257: 9005-9015

Goodman M, Braunitzer G, Kleinschmidt I, Aschauer H (1983) The analysis of a protein polymorphism. Evolution of monomeric and dimeric hemoglobins of Chironomus thummi thummi. Hoppe-Seyler's Z. Physiol. Chem. 364: 205-217

Gotoh T, Kamada Y (1980) Subunit structure of erythrocruorin from the polychaete Tylorrhynchus heterochaetus. Biochem. J. (Tokyo) 87: 557-562

Grinich NP, Terwilliger RC (1980) The quaternary structure of an unusual high molecular weight intracellular hemoglobin from the bivalve mollusc Barbatia reevana. Biochem. J. 189: 1-8

Ilan E, Daniel E (1979a) Structural diversity of arthropod extracellular hemoglobins. Comp. Biochem. Physiol. 63B: 303-308

Ilan E, Daniel E (1979b) Hemoglobin from the tadpole shrimp Lepidurus apus lubbocki. Biochem. J. 183: 325-330

Ilan E, David MM, Daniel E (1981) Erythrocruorin from the crustacean Caenestheria inopinata. Quaternary structure and subunit arrangement. Biochemistry 20: 6190-6194

Ilan E, Daniel E (1982) Erythrocruorin from the water flea Daphnia magna. Quaternary structure and arrangement of subunits. Biochem. J. 207: 197-207

Imamura T, Baldwin TO, Riggs A (1972) The amino acid sequence of the monomer hemoglobin component from the bloodworm Glycera dibranchiata. J. Biol. Chem. 247: 2785-2797

Irie T, Usuki I (1980) Disparity of native oxyhemoglobine components isolated from Paramecium caudatum and P. primaurelia. Comp. Biochem. Physiol. 62B: 549-554

Kapp OH, Ohtsuki M, Crewe AV, Vinogradov SN (1982) Scanning transmission electron microscopy of extracellular invertebrate hemoglobins. Biochim. Biophys. Acta 704: 546-548

Kapp OH, Crewe AV (1984) Comparison of the molecular size and shape of extracellular hemoglobins of Tubifex tubifex and Lumbricus terrestris. Biochim. Biophys. Acta (in press)

Keilin D, Ryley JF (1953) Haemoglobin in Protozoa. Nature 172: 451

Lee DL, Smith MH (1965) Haemoglobin in parasitic animals. Exp. Parasit. 16: 392-424

Messerschmidt U, Wilhelm P, Pilz I, Kapp OH, Vinogradov SN (1983) The molecular size and shape of the extracellular hemoglobin of Nephtys incisa. Biochim. Biophys. Acta 742: 366-373

Mezzasalma V, Di Stefano L, Piazzese S, Zagra M, Ghiretti-Magaldi A, Carbone R, Salvato B (1983) Structural studies on Spirographis spallanzanii chlorocruorin. Life Science Reports, suppl. 1: 187-191

Moens L, Kondo M (1978) Evidence for a dimeric form of Artemia salina extracellular hemoglobin. Eur. J. Biochem. 82: 65-72

Moens L (1982) The extracellular haemoglobins of Artemia sp. A biochemical and onto-genetical study. Academiae Analecta 44: 1-21

Okazaki T, Briehl RW, Wittenberg JB, Wittenberg BA (1965) The haemoglobin of Ascaris perienteric fluid. II. Molecular weight and subunits. Biochim. Biophys. Acta 111: 496-502

Padlan EA, Love WE (1974) Three-dimensional structure of hemoglobin from the bloodworm Glycera dibranchiata. J. Biol. Chem. 127: 309-338

Phelps CF, Antonini E, Brunori M, Kellett G (1972) The kinetics of binding of oxygen and carbon monoxide to Gastrophilus haemoglobin. Biochem. J. 129: 891-896

Phillips JJ (1978) The occurrence and distribution of hemoglobin in the entosymbiotic rhabdocoel Paravortex scrobicularia. Comp. Biochem. Physiol. 61A: 679-683

Pilz I, Schwarz E, Vinogradov SN (1980) Small angle X-ray studies of Lumbricus terrestris hemoglobin. Int. J. Biol. Macromol. 2: 279-283

Royer WE, Braden BC, Jacobs HC, Love WE (1981) Crystals of whole molecules of Lumbricus erythrocruorin. In: Lamy J and Lamy J (eds) Invertebrate Oxygen Binding Proteins, Marcel Dekker, New York, pp. 337-341

Shlom JM, Vinogradov SN (1973) A study of the subunit structure of Lumbricus terrestris hemoglobin. J. Biol. Chem. 248: 7904-7912

Shlom JM, Vinogradov SN (1975) Subunits of Placobdella hemoglobin. Comp. Biochem. Physiol. 51B: 389-392

Steers Jr E, Davis RH (1979) Purification and characterization of myoglobin from Para-mecium tertaurelia. Comp. Biochem. Physiol. 62B: 393-402

Steigemann W, Weber E (1979) Structure of erythrocruorin in different ligand states refined at 1.4 A resolution. J. Mol. Biol. 127: 309-338

Suzuki T, Takagi T, Gotoh T (1982) Amino acid sequence of the smallest polypeptide chain containing heme of extracellular hemoglobin from the polychaete Tylorrhynchus hetero-chaetus. Biochim. Biophys. Acta 708: 253-258

Terwilliger NB, Terwilliger RC, Schabtach E (1976a) The quaternary structure of a mollusc (Helisoma trivolvis) extracellular hemoglobin. Biochim. Biophys. Acta 453: 101-110

Terwilliger RC, Terwilliger NB, Schabtach E (1976b) Comparison of chlorocruorin and annelid hemoglobin quaternary structures. Comp. Biochem. Physiol. 55A: 51-55

Terwilliger RC (1978) The respiratory pigment of the serpulid polychaete Serpula vermi-cularis. Structure of its hemoglobin and chlorocruorin. Comp. Biochem. Physiol. 61B: 463-469

Terwilliger NB, Terwilliger RC (1978) Oxygen binding domains of a marine clam (Cardita borealis) extracellular hemoglobin. Biochim. Biophys. Acta 537: 77-85

Terwilliger RC, Terwilliger NB, Schabtach E (1978) Extracellular hemoglobin of a marine clam (Cardita borealis). Comp. Biochem. Physiol. 59B: 9-14

Terwilliger RC, Terwilliger NB (1983) Oxygen binding domains in invertebrate hemoglobins. Life Chemistry Reports, suppl. 1: 227-238

Tuchsmid PE, Kunz PA, Wilson KJ (1978) Isolation and characterization of the hemoglobin from the lanceolate fluke Dicrocoelium dendriticum. Eur. J. Biochem. 88: 387-394

Vainshtein BK (1981) The structure of leghemoglobin. In: Dodson G, Glusker CJP, Sayre D (eds) Structural Studies of Molecular Biological Interest, Oxford University Press, London, pp. 39-43

van Holde KE, Miller KI (1982) Hemocyanins. Quart. Rev. Biophys. 15: 1-129

Vinogradov SN, Hall BC, Shlom JR (1976) Subunit homology in invertebrate hemoglobins: a primitive heme-binding chain . Comp. Biochem. Physiol. 53B: 89-92

Vinogradov SN, Shlom JM, Hall BC, Kapp OH, Mizukami H (1977) The dissociation of Lumbricus terrestris hemoglobin: a model of its subunit structure. Biochim. Biophys. Acta 492: 136-155

Vinogradov SN, Orii Y (1980) Subunits of Potamilla leptochaeta chlorocruorin. Comp. Biochem. Physiol. 67B: 183-185

Vinogradov SN, Shlom JM, Kapp OH, Frossard P (1980) The dissociation of annelid extracellular hemoglobins and their subunit structure. Comp. Biochem. Physiol. 67B: 1-12

Vinogradov SN, Kapp OH, Ohtsuki M (1982) The extracellular haemoglobins and chlorocruo-rins of annelids. In: Harris JR (ed) Electron Microscopy of Proteins, Vol. 3, Academic Press, London, pp. 135-164

Vinogradov SN, Van Gelderen J, Polidori G, Kapp OH (1983) Dissociation of the extracellular hemoglobin of Nephtys incisa. Comp. Biochem. Physiol. 76B: 207-214

Vinogradov SN, Standley PR, Mainwaring MG, Kapp OH, Crewe AV (1985) Molecular size of Myxicola infundibulum chlorocruorin and of its subunits. Biochim. Biophys. Acta (submitted)

Wittenberg BA, Briehl RW, Wittenberg JB (1965) The haemoglobins of invertebrate tissues. Biochem. J. 96: 363-371

Wood EJ, Mosby LJ (1975) Physicochemical properties of Planorbis corneus erythrocruorin. Biochem. J. 149: 437-445

Wood EJ, Mosby LJ, Robinson MS (1976) Characterization of the extracellular haemoglobin of Haemopsis sanguisuga. Biochem. J. 153: 589-596

Wood EJ, Barker C, Moens L, Jacob W, Heip J, Kondo M (1981) Biophysical characterization of Artemia salina extracellular haemoglobin. Biochem. J. 193: 353-359

Yager TD, Terwilliger NB, Terwilliger RC, Shabtach E, van Holde KE (1982) Organization and physical properties of the giant extracellular hemoglobins of the clam Astarte castanes. Biochim. Biophys. Acta 709: 194-203

Physiological Adaptations and Subunit Diversity in Hemocyanins

J. BONAVENTURA, C. BONAVENTURA

I. INTRODUCTION

One may question the degree to which physiological adaptations of multisubunit oxygen carrying proteins to the requirements of diverse organisms are based on the existence of structurally and functionally distinct types of subunits. This question can be most easily addressed through studies of the hemocyanins, where multiple types of subunits can self-assemble into truly giant molecules with highly developed allosteric properties. A consideration of our present knowledge concerning subunit differences leads rather directly to the conclusion that both the assembly of the high molecular weight aggregates found in vivo and their physiological function may be directly related to the types of subunits present. Much of our present understanding has come from study of the structure, function and assembly of the chelicerate hemocyanins, present in the horseshoe crabs, scorpions, and spiders. Only recently has it been possible to perform comparable experiments with crustacean hemocyanins. One crustacean hemocyanin, that of the lobster Panulirus interruptus, has proven to be a good model system for studies of subunit diversity. We find that in this system the diverse subunits differ in their ability to self-assemble, in their relative sensitivities to calcium and magnesium, and, of particular physiological importance, in the extent to which an organic cofactor can modulate their oxygen binding properties. Studies with hemocyanin of the garden snail, Helix pomatia and of the sea snail, Murex fulvescens, can be cited as evidence for the presence of diverse subunits that contribute significantly to the assembly and function of molluscan hemocyanins.

One feature shared by all hemocyanins is the bis-copper active site where oxygen is reversibly bound. We have undertaken studies of hydrogen peroxide interactions with subunits of varied hemocyanins in order to clarify the extent and significance of active-site heterogeneity between subunits. The hemocyanins of arthropods and molluscs, while similar in physiological function, have long been recognized as differing in regard to their interactions with hydrogen peroxide. We have now found that the ability of hydrogen peroxide to regenerate oxyhemocyanin from the oxidized state varies appreciably between the isolated subunits of Limulus hemocyanin. Subunits III and IV are diametrically opposed. Subunit III exhibits "mollusc-like" character while subunit IV more closely resembles the "typical" arthropod hemocyanin that becomes completely oxidized in the presence of hydrogen peroxide. These results carry interesting implications with regard to active-site heterogeneity among the hemocyanins and may provide an explanation for a number of previously perplexing observations such as the incomplete removal of copper from a number of hemocyanins that accompanies treatment with mercury.

When respiratory oxygen demands exceed the rate of oxygen delivery that can be accomplished by diffusional processes, the job of transporting oxygen from the environment to the respiring tissues is done by hemoglobins, hemocyanins, or hemerythrins. These three classes of oxygen-carrying proteins bear little resemblance to one another in their molecular architectures, but play the same physiological role. Since they typically occur in high concentrations in vivo and are relatively easy to isolate, they have been favored for biochemical studies of structure-function relationships. We suggest that these proteins are excellent model systems for studies directed toward gaining a further insight into how specialized subunits confer properties of physiological advantage upon multisubunit assemblages.

The three classes of oxygen-carrying proteins vary widely in their functional properties, allowing for an elegant fit between their functional role as oxygen carriers and the diverse physiological and environmental parameters encountered by a very wide range of air-breathing and water-breathing organisms. The extent to which the existence of diverse types of subunits are responsible for this functional flexibility can perhaps be best addressed through studies of the hemocyanins. These oligomers typically contain multiple types of subunits and synthetic homo- and heteropolymers can be prepared. In the following, we will briefly address our state of knowledge with respect to the physiological significance of subunit diversity in the hemocyanins and present the results of some recent studies that bear on this topic.

General information on the hemocyanins and their physical and functional properties can be found in several review articles (Bonaventura and Bonaventura, 1983; Ellerton et al., 1983; van Holde and Miller, 1982). In brief, the hemocyanins are high molecular weight copper proteins found in the hemolymph of gastropods and cephalopods of the phylum Mollusca and of xiphosaurans, arachnids, and crustacea of the phylum Arthropoda. They are not contained within cells and it has been postulated that aggregation into giant molecules facilitates the maintenance of osmotic equilibrium (Snyder and Mangum, 1982). Aggregation allows for highly developed allosteric properties: cooperativity between oxygen binding sites and pH and ion effects on the oxygen binding process are common features (Bonaventura and Bonaventura, 1981). Cooperative interactions between oxygen binding sites (homotropic interactions) arise by virtue of conformational transitions between conformational states of high affinity (R-states) and of low affinity (T-states). The switchover between conformational states in the oxygenation pathway can be greatly influenced by interactions with effectors that bind at positions distant from the active site (heterotropic interactions). The basis for heterotropic interactions is a preferential binding of the effector molecules to a given conformational state. Thus the influence of effectors can be to activate or to inhibit a protein system. For example, in respiratory proteins with normal Bohr effects, protons stabilize the low affinity conformation. Protons are liberated as the conformational transition occurs upon oxygenation. The reverse situation can occur, and in hemocyanins the heterotropic effects of protons can stabilize either high or low affinity conformers.

The hemocyanins first came to the attention of biochemists and physiologists by virtue of their high molecular weights and their distinctive characteristic of reversible assembly and disassembly, documented in some of the pioneering work of Svedberg and colleagues in the early days of ultracentrifugation (Svedberg and Hedenius, 1934; Svedberg and Petersen, 1940). As in the case of oxygen binding, the state of aggregation of these molecules is highly dependent on such parameters as pH, the presence or absence of oxygen or divalent cations, temperature, and protein concentration (Bonaventura and Bonaventura, 1983; Ellerton et al., 1983; van Holde and Miller, 1982). The allosteric nature of hemocyanins is clearly an adaptive feature. In the following, we consider the extent to which subunit diversity may contribute to the control mechanisms whereby these proteins are able to respond to variable physiological requirements.

II. SUBUNIT HETEROGENEITY IN MOLLUSCAN HEMOCYANINS

Molecular heterogeneity, as it relates to dissociation and reassembly of the molluscan hemocyanins, was noted at a fairly early stage. Researchers in many laboratories adopted the methods of Konings et al. (1969) for separation of α and β types of hemocyanins from the edible snail Helix pomatia, the two components differing in salt stability. Most studies with Helix hemocyanin have been done after separation of these two components. Subunits of the molluscan hemocyanins are very large and complex. In gastropod molluscs, the subunits have molecular weights of 350,000-400,000, corresponding to 1/20 of the undissoci- ated molecule (Brouwer and Kuiper, 1973; Siezen and van Bruggen, 1974). Electron micrographs of the subunits show a "string of beads" structure with 7-8 flexibly connected globules (Siezen and van Bruggen, 1974). Studies of the subunits have been complicated by proteolytic cleavage into smaller fragments, usually in multiples of 50,000 with one oxygen- binding site per 50,000 dalton domain (Bannister et al., 1977, Bonaventura et al., 1977; Bonaventura and Bonaventura, 1983; Brouwer, 1975; Brouwer et al., 1976; Gielens et al., 1975; Gielens et al., 1977; Lontie and Witters, 1973). Subunits of H. pomatia α-hemocyanin seem to be heterogeneous in electrophoretic mobility, as judged by the reversible boundary spreading test (Siezen and van Driel, 1973). The presence of two different polypeptide chains in this hemocyanin was suggested, based upon the interpretation of the time course of the proteolytic digestion of the native molecule (Brouwer and Kuiper, 1973). Until studies were done with hemocyanin of the sea snail, Murex fulvescens, there was, however, no conclusive evidence regarding the existence of more than one type of subunit in molluscan hemocya- nins.

The subunits of M. fulvescens hemocyanin can be fractionated into two components which migrate as single bands on regular disc gels (Brouwer et al., 1978). The native 100S molecule is composed of approximately equimolar quantities of these subunits. It is significant that both subunits are required for reassembly of the 100S molecule from its 11S subunits. The isolated subunits show little allosteric control, while the reassociated hemocyanin exhibits cooperative interactions and pH dependence of oxygen binding equilibria and kinetics which are very similar to those of the undissociated molecules found in the hemolymph. Amino acid composition, neutral hexose content, and functional behavior of the two subunits were found to be very similar. The equilibria and kinetics of oxygen binding by both subunits are heterogeneous. Heterogeneity in this hemocyanin is therefore not only found on the subunit level, but on the domain level as well (Brouwer et al., 1978).

Most of the structural and functional studies on gastropod hemocyanins have been done with the pigments of modern univalves (order Caenogastropoda) or with the snails belonging to the gastropod subclass of Pulmonata. We considered that knowledge of structural and functional characteristics of hemocyanin isolated from a primitive mollusc might lead to an interesting comparison with the properties of other molluscan hemocyanins. Of these, that of the keyhole limpet, Megathura crenulata was selected for study, in part because of its frequent use as a "typical" molluscan hemocyanin in immunological investigations. In the hemolymph, as well as in buffers containing calcium, Megathura hemocyanin exists in five states of aggregation (Senozan et al., 1981). These states have sedimentation coefficients of 102S, 130S, 150S, 170S, and 186S. Other marine gastropods, like M. fulvescens, have hemocyanins whose predominant species under physiological conditions is a molecule with a sedimentation coefficient of 100S. The 100S molecule typically dissociates when divalent cations are removed from the suspending medium. This is not the case for Megathura hemocyanin, which apparently has a higher affinity for the cations that stabilize its structure. Moreover, this hemocyanin, unlike that of other marine gastropods, exhibits a normal (positive) Bohr effect. Its pH sensitivity and cooperativity in oxygen binding persists after dialysis against buffers devoid of divalent cations. Experiments with dissociated Megathura hemocyanin analogous to those cited above for Murex hemocyanin, revealed that the oligomer is composed of two distinct kinds of subunits (Senozan et al., 1981). The stoichiometry of the two types of subunits was determined to be approximately 1:2, in contrast to the 1:1 stoichiometry found for the two types of Murex hemocyanin subunits. Studies in progress are directed toward evaluation of whether the unequal stoichiometry observed in the keyhole limpet hemocyanin is correlated with the formation of extended aggregates (greater than 100S) that is one of the distinguishing features of this hemocyanin.

Within the phylum Mollusca, the major classes appear to have somewhat distinctive hemocyanins. In the cephalopods, the hemocyanins found in the blood appear as cylindrical molecules with approximately the same diameter, but only half the height, of those found in the blood of gastropods. This feature is correlated with a lower sedimentation coefficient for the native molecule, giving values of 58-59S in contract to values of 100S of most gastropod hemocyanins and values of 64-66S for half-dissociated gastropod hemocyanins. Several studies concerning the structure and function of these molecules have been reported (Bonaventura and Bonaventura, 1983; Ellerton et al., 1983; van Holde and Miller, 1982). Studies were undertaken to address the question of whether cephalopod hemocyanins also contain multiple subunits. The results of a study of Nautilus pompilius hemocyanin clearly showed that more than one kind of high-molecular-weight polypeptide chain is involved in its assembly (Bonaventura et al., 1981). This report of multiple subunit types in a cephalopod hemocyanin makes it seem probable that subunit diversity in molluscan hemocyanin may be the rule rather than the exception. The stoichiometry of diverse subunit types in Nautilus hemocyanin, as determined by integrating the chromatographic peaks, suggests that if the 58S molecule is a mixed decamer, there are three A-, five B-, and two C-type subunits present. Components A and C are quite similar in function and together they account for 50% of the mixture. The possibility was not excluded that component C is a degradation product of component A. Moreover, the experiments do not rule out the existence of a mixture of different decamers (Bonaventura et al., 1981).

III. HETEROGENEITY IN ARTHROPODAN HEMOCYANINS

The hexamers, dodecamers and higher ordered aggregates characteristic of the arthropod hemocyanins vary appreciably in molecular architecture from the one- or two-tiered cylinders typical of the molluscan hemocyanins. Subunit diversity, however, is widespread in the arthropod hemocyanins as well. In fact, all native arthropod hemocyanins thus far subjected to detailed study have shown indications of subunit heterogeneity. A systematic study of electrophoretic heterogeneity in arthropod hemocyanins has been published (Markl et al., 1979). The less complex nature of the arthropod hemocyanin subunits, where each subunit has a molecular weight of about 75,000 and contains a single oxygen binding site, makes these molecules more amenable to study than the subunits of the molluscan hemocyanins. The structurally diverse subunits can be used as subtle probes of the structural origins of specific protein-protein interactions involved in self-assembly and in allosteric modulation of oxygen affinity.

The hemocyanin of the horseshoe crab Limulus polyphemus is characteristic of arthropod hemocyanins in that it is a high-molecular-weight oligomer composed of functionally and structurally distinct subunits. The protein forms a 48-subunit complex, the largest form of arthropod hemocyanin, whose oxygen-binding characteristics are modulated by subunit interactions within the oligomer (Bonaventura et al., 1974; Sullivan et al., 1974, Sullivan et al., 1976). Differences exist in the oxygen-binding affinity and modulation by chloride ions for the isolated Limulus hemocyanin subunits (Bonaventura et al., 1974; Bonaventura and Bonaventura, 1981; Brenowitz et al., 1981; Brenowitz et al., 1984; Sullivan et al., 1974; Sullivan et al., 1976). Similarly, Bijlholt et al. (1979) have shown that omission of specific chromatographic zones prevents reassembly of the 48-subunit structure, with the aggregation state reached dependent on which zone is omitted. These results are summarized in Tables 1 and 2.

TABLE 1. Structural Role of <u>Limulus</u> Hemocyanin Fractions in the Assembly of the 3×10^6 dalton 48-Mer[*]

Chromatographic zone	Putative role in assembly	
I	No unique role	
II	Dodecamer	24-mer
III	24-mer	48-mer
IV	24-mer	48-mer
V	Hexamer	dodecamer

[*]From Brenowitz <u>et al.</u> (1981)

TABLE 2. Oxygen Affinity of <u>Limulus</u> hemocyanin components, partially purified by ion-exchange chromatography[*]

Limulus zone	$P_{\frac{1}{2}}$ at pH 7.5 (mm Hg)	k at pH 7.5 (sec^{-1})	$P_{\frac{1}{2}}$ at pH 9.0 (mm Hg)
I	2.1	7.5	2.2
II	1.4	7.14	1.62
III	1.4	4.0	1.5
IV	1.0	2.4	1.0
V	5.9	33.3	5.7

[*]Data were collected at 20 °C in 0.05 M Tris, pH 7.5 and 9.0. Data from Bonaventura <u>et al.</u> (1974).

The evidence so far accumulated indicates that the subunit heterogeneity in <u>Limulus</u> hemocyanin is the result of differences in the polypeptides at the level of their primary structure. Tryptic peptide maps, limited proteolysis by formic acid hydrolysis, CNBr cleavage and tryptic digestion of the isolated subunits of chelicerate hemocyanins all point to the uniqueness of the different polypeptide chains (Jollès <u>et al.</u>, 1981; Sullivan <u>et al.</u>, 1976). Sequence data, although limited, provides convincing additional evidence of this. Table 2 shows the oxygen affinities of the five major chromatographic fractions. Table 2 also shows the first order oxygen dissociation constant (k) for these fractions. It can be seen that there is an inverse relationship between the oxygen affinity and the value of k. This illustrates a common phenomenon in hemocyanins, that is, that the rate at which oxygen dissociates from the oxyhemocyanin molecule is highly correlated with its overall oxygen affinity (Bonaventura et al., 1977, Brunori et al., 1977). Comparison of the $P_{\frac{1}{2}}$ values of Table 2 at pH 7.5 and 9.0 also shows that the individual subunits do not show any appreciable Bohr effect in the absence of effectors, in contrast to the whole molecule under physiological conditions.

The major chromatographic zones represented in Tables 1 and 2 do not fully represent the extent of subunit diversity in Limulus hemocyanin. Using polyacrylamide electrophoresis, Markl et al. (1979) observed 12 bands; while using immunoelectrophoresis, Hoylaerts et al. (1979) and Lamy et al. (1979) observed 8 subunits. Refinement of the chromatographic separation procedures has led to the isolation of 8 immunologically distinct subunits as well as additional charge isomers which cannot be distinguished immunologically. Alkaline electrophoresis revealed 15 bands and isoelectric focusing up to 17 (Brenowitz et al., 1981). On the basis of extensive control experiments, including composit acrylamide-agarose immunoelectrophoresis and checks for conformational isomers, aggregation, proteolysis, and other types of degradation, we concluded that the electrophoretic heterogeneity of immunologically identical subunits is not artifactual (Brenowitz et al., 1981). However, if a charge substitution is not in a critical location, we would expect the electrophoretically distinct but immunologically identical subunits to have identical assembly roles. Experiments to date indicate this to be the case. As mentioned, reassociation studies with partially purified subunit preparations of Limulus subunits have shown that the different subunits play different roles in stabilizing the native molecule. Similar experiments, with immunologically and electrophoretically pure subunits, confirm the varied roles of specific subunits in assembly and also show that immunologically identical, yet electrophoretically different, proteins can substitute for each other in assembly (Brenowitz et al., 1984). Correspondingly, distinctly different assembly characteristics are observed among the immunologically nonidentical subunits. The same pattern of the immunological identity of a subunit mirroring its structural and functional properties is seen in the affinity and allosteric modulation of the oxygen-binding properties of the subunits.

Studies like those summarized above for Limulus hemocyanin have been conducted with hemocyanins of two related species, the scorpion, Androctonus australis (Lamy et al., 1980) and the tarantula, Eurypelma californicum (Markl et al., 1981). It was found in all of these chelicerate hemocyanins that subunits distinguished as "hexamer-formers" and "linkers" were needed to make "multi-hexameric" aggregates (van Bruggen et al., 1980). Hybrids could be formed by artificial mixtures of these under appropriate assembly conditions (van Bruggen et al., 1980).

In adaptations to the frequently changing status of their environmental niches, organisms must take "use" of the allosteric nature of their oxygen binding proteins. Studies of the arthropod hemocyanins have shown that oxygen affinity can be altered by stabilization of Low affinity "T" structures or stabilization of High affinity "R" structures. Heterotropic effects may thus be complementary or antagonistic (agonistic-antagonistic).

For example, in the Limulus hemocyanin system, the binding of chloride favors a low affinity state, whereas the binding of protons favors a high affinity state (Bonaventura et al., 1974, Brenowitz et al., 1981; Brenowitz et al., 1984; Sullivan et al., 1974; Sullivan et al., 1976). The Bohr effects of hemocyanins can be positive or negative and no structural explanation has yet been advanced to account for this. The negative Bohr effect of Limulus hemocyanin is not a common feature of hemocyanins although it is found in a number of other hemocyanin molecules. The different subunits of the Limulus oligomer probably contribute unequally to its pH sensitivity, since pH effects on oxygen binding by the isolated subunits are known to differ (Bonaventura et al., 1974; Bonaventura and Bonaventura, 1981; Brenowitz et al., 1981; Brenowitz et al., 1984; Sullivan et al., 1974; Sullivan et al., 1976). Chloride effects on the Limulus oligomer may also be associated with specific subunit types. The effect of chloride on Limulus II hemocyanin is almost as strong as it is on the 60S oligomer of Limulus hemocyanin. Limulus IV, on the other hand, has essentially no sensitivity to chloride (Bonaventura et al., 1974; Bonaventura and Bonaventura, 1981, Brenowitz et al., 1981; Brenowitz et al., 1984; Sullivan et al., 1974; Sullivan et al., 1976). Studies with hybrid molecules with subunits from species with differing pH and chloride sensitivities may further clarify the roles of tertiary and quaternary changes in these heterotropic interactions.

Dissociated crustacean hemocyanins, like those of the chelicerate hemocyanins discussed above, typically show multiple electrophoretic bands. Structural and functional analysis of these subunits has been handicapped by technical difficulties in their isolation. Considerable progress has been made, however, in isolation of distinct subunits in a few crustacean hemocyanins, most notably in case of hemocyanin of the lobster Panulirus interruptus (van Eerd and Folkerts, 1981). The three-dimensional structure of a subunit of this hemocyanin has recently been determined to a resolution of 3.2 Å (Gaykema et al., 1984). In the following, we will present some recently acquired data that points to the physiological significance of subunit diversity in this system.

The oligomeric assemblages of crustacean hemocyanins are typically cooperative and responsive to the presence of small effector molecules. In contrast, the subunits which comprise such systems are, when analyzed in a monomeric state, noncooperative and show little or reduced sensitivity to allosteric effectors. We note, however, that homohexamers of isolated subunits can, under some conditions, exhibit allosteric behavior (Brenowitz, 1982). It is clear that the assembly of subunits into the oligomer is intimately associated with the maintenance of the physiological function of the oligomer found in vivo. The assembly can alter the oxygen binding properties of the subunits toward higher or lower oxygen affinity. A comparison of the subunits and native hemocyanins of the Atlantic and Pacific lobsters provide an excellent illustration of this phenomenon. The native molecules found in Homarus americanus, the Atlantic lobster, and Panulirus interruptus, the Pacific lobster, are very close to one another in their oxygen binding characteristics under physilogical conditions. Both are cooperative, with Hill coefficients greater than 3. Dissociation of the oligomers into their subunits, by high pH treatment and removal of divalent cations, results in molecules with widely differing oxygen binding curves. In the case of H. americanus, the isolated subunits are of relatively low affinity. The isolated subunits of Panulirus, on the other hand, are of relatively high oxygen affinity. Both types of subunits show little or no cooperativity in the absence of allosteric effectors, but for the unfractionated Homarus subunits the Log P_{50} at pH 7.0 is above 2.0, while for Panulirus subunits under the same conditions the Log P_{50} value is about 1.0. One conclusion to be drawn from these studies is that the behavior of subunits in an isolated state is not always predictive with respect to their behavior when assembled into oligomers.

We have carried out detailed studies on the assembly and oxygen binding character-istics of the isolated subunits of the Panulirus hemocyanin system. Significant differences exist between the absolute affinities and Bohr effects of the isolated subunits. Moreover, addition of calcium or magnesium affects the different subunits to variable extents in regards to their assembly and oxygen binding characteristics. Of particular physiological interest is the fact that homohexamers formed by the different subunits of Panulirus hemocyanin show differences in their ability to interact with L-lactate. This metabolite is a specific effector of hemocyanin function (Johnson et al., 1984; Mangum, 1983). Only two of the three Panulirus subunits studied in detail were found to be capable of forming homohexamers that showed a significant response to L-lactate (Johnson, 1984). The subunits that show lactate sensitivity are less closely related to each other, as determined by immunological techniques, than either are to the lactate-insensitive subunit (Folkerts and van Eerd, 1981). This suggests that sensitivity to lactate was an ancestral state that was subsequently lost by some of the subunits (Johnson, 1984).

A homohexamer whose oxygen affinity is sensitive to the presence of L-lactate might be expected to have six lactate binding sites. However, analysis of the relevant oxygen-binding curves using an oxygen-linkage equation indicates that both the native hexamers (containing several types of subunits) and lactate-sensitive homohexamers (prepared from isolated subunits) possess only one lactate binding site per hexamer (Johnson, 1984). The lack of a 1:1 ratio of lactate binding sites to subunits indicates that the lactate binding site occurs between subunits of the assembled hexameric state. It is thus apparent that the expression of lactate-sensitivity in Panulirus hemocyanin depends on the formation of a specific site within the quaternary structure of the protein (Johnson, 1984).

IV. DIVERSITY IN ACTIVE SITES OF HEMOCYANIN SUBUNITS

Another question of physiological and biochemical interest that may be addressed with hemocyanins concerns the extent to which their subunits show active-site heterogeneity. The active sites of all hemocyanins contain two copper atoms that are linked directly to the amino acid residues of the protein. Oxygen is thought to bridge across these two coppers in a μ-dioxo fashion. Examination of oxygen binding by hemocyanins reveals extensive variability in the affinity of these molecules for oxygen and in the extent to which site-site interactions and cofactor effects can alter the affinity of the active site for oxygen. Carbon monoxide can also bind at the active site, but in this case cofactor effects and cooperative interactions between binding sites are reduced or absent (Bonaventura et al., 1974). The variable oxygen and carbon monoxide affinities shown by different hemocyanins provides evidence that the electronic environment of the bis-copper active site varies from species to species. Functional differences that have been shown to exist between the isolated subunits of a given species, in terms of oxygen and carbon-monoxide binding, indicates that active site heterogeneity exists at the subunit level as well. In the following, the question we address concerns the basis of a major distinction drawn between the active sites of the mollusc and arthropod hemocyanins.

Differences between the active sites of the mollusc and arthropod hemocyanins have long been recognized, with the predominant functional distinction being the extent that they become irreversibly oxidized by hydrogen peroxide. In the molluscan hemocyanins, oxidative processes at the active site are initiated upon peroxide addition, but the methemocyanin form that is generated is capable of further interaction with peroxide that returns the hemocyanin to the oxygenated state and, in a cyclic fashion, eventually depletes the hydrogen peroxide concentration in solution. In contrast, the interaction of peroxide with arthropod hemocyanins typically leads, irreversibly, to oxidized active sites. A number of papers have been published that concern this effect and related phenomena. Of these, readers may find particularly useful the papers by Felsenfeld and Printz (1959), Lontie et al. (1982), Solomon et al. (1982), and Witters et al. (1974) that are relevant and also present rather extensive bibliographies.

The regeneration of oxyhemocyanin from oxidized (met) hemocyanin by peroxide appears to require a good "fit" of peroxide to the geometry of the oxidized bis-copper site. This condition is satisfied in the mollusc hemocyanins, but not in the arthropod hemocyanins where the site "...appears to be somewhat distorted, which strongly affects exogenous ligand binding to the binuclear copper site and, by extension, the peroxide regeneration of the met form to oxy" (Solomon et al., 1982). Spectroscopic studies, which led to the conclusion quoted above, placed Limulus hemocyanin in a position intermediate between the hemocyanins of molluscs and other arthropods (Solomon et al., 1982). We reviewed the published data on the interactions of Limulus hemocyanin with peroxide and found, significantly, that a small fraction of the protein was not oxidized even at light peroxide concentrations (Fig. 2, Felsenfeld and Printz, 1959). With the suspicion that this might relate to subunit heterogeneity, we undertook an analysis of peroxide interactions with isolated Limulus subunits. As shown in Table 3, specific Limulus hemocyanin subunits vary appreciably in their ability to regain the oxy state after oxidation by peroxide. Some, like typical arthropod hemocyanins, lose the 340 nm copper-oxygen absorbance band upon exposure to peroxide while others, like mollusc hemocyanins, do not. These results may be the key to understanding why mercury treatment of Limulus hemocyanin results in an incomplete inactivation of the active site

(Brouwer et al., 1983). In both peroxide and mercury interactions with hemocyanins the geometry at the active site may be constrained by the presence or absence of an internal (endogenous) bridge between the two copper atoms. An internal bridge was postulated on the basis of spectral analysis of various active-site derivatives, most of which was done using hemocyanin of the mollusc, Busycon (Solomon et al., 1982). The recent X-ray analysis of a hemocyanin subunit of the arthropod Panulirus does not indicate the presence of an endogenous bridge (Gaykema et al., 1984). By determining that the Limulus subunits can have either mollusc- or arthropod-like character in regard to peroxide regeneration of oxy forms, the structural comparisons needed to clarify the underlying mechanisms may be greatly facilitated. There is no doubt that structural comparisons between subunits of the same species may more readily identify the significant factors involved than was previously possible when comparisons had to be drawn between such architecturally distinct molecules as the mollusc and arthropod hemocyanins.

TABLE 3. Percentage of oxyhemocyanin at varied times after peroxide addition

Samples of purified, air-equilibrated subunits III and IV of Limulus hemocyanin of equal concentration were examined at various times after addition of a one hundred-fold molar excess of hydrogen peroxide. The percentage of oxyhemocyanin, based on the 340 nm absorbance, varied as shown above, indicating that peroxide interactions with subunit IV led to extensive active-site oxidation, while for subunit III the initial oxidative attack was followed by a regeneration of the oxy form. Subunits were isolated as described elsewhere (Brenowitz, 1982) and monitored in 0.05 M Tris buffer, pH 8.9, 20 °C

	5 min.	5 h.
Limulus Hemocyanin Subunit III	60%	100%
Limulus Hemocyanin Subunit IV	60%	5%

ACKNOWLEDGMENTS

The work reported in this paper was funded by grant ESO 1908 from the National Institutes of Health, grant PCM 8309857 from the National Science Foundation, and grant N00014-83-K-0016 from the United States Office of Naval Research. We also wish to acknowledge our friends and colleagues in Louvain, Groningen, Munich, Rome and Tours whose insightful studies of structure-function relationships in hemocyanins have contributed significantly to the concepts presented here.

REFERENCES

Bannister JV, Galdes A, Bannister WH (1977) The oxygen equilibrium of Murex trunculus haemocyanin. In: Bannister JV (ed) Structure and Function of Haemocyanin. Springer-Verlag, Berlin, New York, pp. 193-205

Bijlholt MMC, van Bruggen EFJ, Bonaventura J (1979) Dissociation and reassembly of Limulus polyphemus hemocyanin. Eur. J. Biochem. 95: 399-405

Bonaventura C, Bonaventura J (1981) Kinetics of oxygen binding by hemocyanins. In: Lamy J, Lamy J (eds) Invertebrate Oxygen Binding Proteins: Structure, Active Site, and Function. Marcel Dekker, New York, pp. 693-701

Bonaventura C, Bonaventura J (1983) Respiratory pigments: structure and function. In: Wilbur K (ed) The Mollusca, Vol. 2, Academic Press, New York, pp. 1-50

Bonaventura C, Sullivan B, Bonaventura J, Bourne S (1974) CO binding by hemocyanins of Limulus polyphemus, Busycon carica, and Callinectes sapidus. Biochemistry 13: 4784-4789

Bonaventura J, Bonaventura C, Sullivan B (1977) Properties of the oxygen-binding domains isolated from subtilisin digests of six molluscan hemocyanins. In: Bannister JV (ed) Structure and Function of Haemocyanin. Springer-Verlag, Berlin, New York, pp. 206-216.

Bonaventura C, Sullivan B, Bonaventura J, Brunori M (1977) Hemocyanin of the horseshoe crab, Limulus polyphemus. A temperature-jump study of the oxygen kinetics of the isolated components. In: Bannister JV (ed) Structure and Function of Haemocyanin. Springer-Verlag, Berlin, New York, pp. 265-270

Bonaventura C, Bonaventura J, Miller KI, van Holde KE (1981) Hemocyanin of the chambered Nautilus: structure-function relationships. Arch. Biochem. Biophys. 211: 589-598

Brenowitz M (1982) The Role of Structurally Distinct Subunits in the Function and Assembly of Limulus polyphemus Hemocyanin, PhD Thesis, Duke University

Brenowitz M, Bonaventura C, Bonaventura J, Gianazza E (1981) Subunit composition of a high molecular weight oligomer: Limulus polyphemus hemocyanin. Arch. Biochem. Biophys. 210 (2): 748-761

Brenowitz M, Bonaventura C, Bonaventura J (1984) Self-association and oxygen-binding characteristics of the isolated subunits of Limulus polyphemus hemocyanin. Arch. Biochem. Biophys. 230 (1): 238-249

Brouwer M (1975) Structural Domains in Helix pomatia Alpha Hemocyanin, Ph.D Thesis, University of Groningen

Brouwer M, Kuiper HA (1973) Molecular-weight analysis of Helix pomatia α-hemocyanin in guanidine hydrochloride, urea and sodium dodecyl-sulfate. Eur. J. Biochem. 35: 428-435

Brouwer M, Wolters M, van Bruggen EFJ (1976) Proteolytic fragmentation of Helix pomatia alpha hemocyanin: structural domains in the polypeptide chain. Biochemistry 15: 18-23

Brouwer M, Ryan M, Bonaventura J, Bonaventura C (1978) Functional and structural properties of Murex fulvescens hemocyanin: isolation of two different subunits required for reassociation of a molluscan hemocyanin. Biochemistry 17: 2810-2815

Brouwer M, Bonaventura C, Bonaventura J (1983) Metal ion interactions with Limulus polyphemus and Callinectes sapidus hemocyanin: stoichiometry and structural and functional consequences of calcium (II), cadmium (II), zinc (II), and mercury (II) binding. Biochemistry 22: 4713-4723

Brunori M, Kuiper HA, Antonini E, Bonaventura C, Bonaventura J (1977) Kinetics of oxygen binding by hemocyanins. In: Lamy J, Lamy J (eds) Invertebrate Oxygen-Binding Proteins: Structure, Active Site, and Function. Marcel Dekker, New York, pp. 693-702

Ellerton HD, Ellerton NF, Robinson HA (1983) Hemocyanin - A current perspective. Progr. Biophys. Molec. Biol. 41: 143-248

Felsenfeld G, Printz MP (1959) Specific reactions of hydrogen peroxide with the active site of hemocyanin. The formation of "methemocyanin". J. Amer. Chem. Soc. 81: 6259-6264

Folkerts A, van Eerd JP (1981) Immunological relatedness of five hemocyanin subunits from the spiny lobster Panulirus interruptus. In: Lamy J, Lamy J (eds) Invertebrate Oxygen Binding Proteins: Structure, Active, Site, and Function. Marcel Dekker, New York, pp. 215-225

Gaykema WPJ, Hol WGJ, Vereijken JM, Soeter NM, Bak HJ, Beintema JJ (1984) 3.2 Å structure of the copper-containing, oxygen-carrying protein Panulirus interrruptus haemocyanin. Nature 309: 23-29

Gielens C, Préaux G, Lontie R (1975) Limited trypsinolysis of β-haemocyanin of Helix pomatia: Characterization of the fragments and heterogeneity of the copper groups by circular dichroism. Eur. J. Biochem. 60: 271-280

Gielens C, Préaux G, Lontie R (1977) Structural investigations of β-haemocyanin of Helix pomatia by limited proteolysis. In: Bannister JV (ed) Structure and Function of Haemocyanin. Springer-Verlag, Berlin, New York, pp. 85-94

Hoylaerts M, Préaux G, Witters R, Lontie R (1979) Immunological heterogeneity of the subunits of Limulus polyphemus hemocyanin. Arch. Intern. Physiol. Biochem. 87: 417-418

Johnson B (1984) Allosteric Interactions of L-Lactate and Inorganic Ions with structurally Distinct Crustacean hemocyanins, Ph.D Thesis, Duke University

Johnson B, Bonaventura C, Bonaventura J (1984) Allosteric modulation of Callinectes sapidus hemocyanin by binding of L-lactate. Biochemistry 23: 872-878

Jollès J, Jollès P, Lamy J, Lamy J (1981) N-terminal sequences and antigenic purity of isolated subunits from Androctonus australis hemocyanin. In: Lamy J, Lamy J (eds) Invertebrate Oxygen-Binding Proteins: Structure, Active Site, and Function. Marcel Dekker, New York, pp. 305-310

Konings WN, Dijk J, Wichertjes T, Beuvery EC, Gruber M (1969) Structure and properties of hemocyanins. IV. Dissociation of Helix pomatia hemocyanin by succinylation into functional subunits. Biochim. Biophys. Acta 188: 43-54

Lamy J, Lamy J, Weill J, Bonaventura J, Bonaventura C, Brenowitz M (1979) Immunological correlates between the multiple subunits of Limulus polyphemus and Tachypleus tridentatus. Arch. Biochem. Biophys. 196: 324-339

Lamy J, Lamy J, Bonaventura J, Bonaventura C (1980) Structure, function and assembly in the hemocyanin system of the scorpion Androctonus australis. Biochemistry 19: 3033-3039

Lontie R, Witters R (1973) Hemocyanin. In: Eichorn GL (ed) Inorganic Biochemistry. Elsevier, Amsterdam, pp. 344-358

Lontie R, Gielens C, Groeseneken D, Verplaetse J, Witters R (1982) Comparison of the active sites of molluscan and arthropodan hemocyanins. In: King TE, Mason HS, Morrison M (eds) Oxidases and related Redox Systems. Pergamon Press, Oxford, New York, pp. 245-261

Mangum CP (1983) On the distribution of lactate sensitivity among the hemocyanins. Mar. Biol. Lett. 4: 139-150

Markl J, Hofer A, Bauer G, Markl A, Kempter B, Brenzinger M, Linzen B (1979) Subunit heterogeneity in arthropod hemocyanins. II. Crustacea. J. Comp. Physiol. 133: 167-175

Markl J, Decker H, Savel A, Linzen B (1981) Homogeneity, subunit heterogeneity, and quaternary structure of Eurypelma hemocyanin. In: Lamy J, Lamy J (eds) Invertebrate Oxygen-Binding Proteins: Structure, Active Site, and Function. Marcel Dekker, New York, pp. 445-454

Senozan NM, Landrum J, Bonaventura J, Bonaventura C (1981) Hemocyanin of the giant keyhole limpet, Megathura crenulata. In: Lamy J, Lamy J (eds) Invertebate Oxygen-Binding Proteins: Structure, Active Site, and Function. Marcel Dekker, New York, pp. 703-717

Siezen RJ, van Driel R (1973) Structure and properties of hemocyanins. VIII. Microheterogeneity of α-hemocyanin of Helix pomatia. Biochim. Biophys. Acta 295: 131-139

Siezen RJ, van Bruggen EFJ (1974) Structure and properties of hemocyanin. XII. Electron microscopy of Helix pomatia α-hemocyanins quaternary structure. J. Mol. Biol. 90: 77-89

Snyder GK, Mangum CP (1982) The relationship between the capacity for oxygen transport, size, shape, and aggregation state of an extracellular oxygen carrier. In: Bonaventura J, Bonaventura C, Tesh S (eds) Physiology and Biology of Horseshoe Crabs. A.R. Liss Inc., New York, pp. 173-188

Solomon EI, Eickman NC, Himmelwright RS, Hwang YT, Plon SE, Wilcox DE (1982) The nature of the binuclear copper site in Limulus and other hemocyanins. In: Bonaventura J, Bonaventura C, Tesh S (eds) Physiology and Biology of Horseshoe Crabs. A.R. Liss Inc., New York, pp. 189-230

Sullivan B, Bonaventura J, Bonaventura C (1974) Functional difference in the multiple hemocyanins of the horseshoe crab, Limulus polyphemus L. Proc. Natl. Acad. Sci. USA 71: 2558-2562

Sullivan B, Bonaventura J, Bonaventura C, Godette G (1976) Hemocyanin of the horseshoe crab Limulus polyphemus. Structural differentiation of the isolated components. J. Biol. Chem. 251: 7644-7648

Svedberg T, Hedenius A (1934) The sedimentation constants of the respiratory proteins. Biol. Bull. (Woods Hole, Mass.) 66: 191-223

Svedberg T, Petersen KO (1940) The Ultracentrifuge. Oxford University Press, London and New York

van Bruggen EFJ, Bijlholt MMC, Schutter WG, Wichertjes T, Bonaventura J, Bonaventura C, Lamy J, Lamy J, Leclerc M, Schneider HJ, Markl J, Linzen B (1980) The role of structurally diverse subunits in the assembly of three cheliceratan hemocyanins. FEBS Lett. 116: 207-210

van Eerd JP, Folkerts A (1981) Isolation and characterization of five subunits of the hemocyanin from the spiny lobster Panulirus interruptus. In: Lamy J, Lamy J (eds) Invertebrate Oxygen-Binding Proteins: Structure, Active Site, and Function. Marcel Dekker, New York, pp. 139-149

van Holde KE, Miller KI (1982) Hemocyanins. Quart. Rev. Biophys. 15: 1-129

Witters R, van Rossen-Usé L, Lontie R (1974) The regeneration with cysteine of gastropod haemocyanin mediated by hydrogen peroxide. Arch. Intern. Physiol. Biochim. 82: 917-924

Functions and Functioning of Crustacean Hemocyanin

B. McMAHON

I. INTRODUCTION

Although the oxygen binding properties of hemocyanin have been known for over 100 years the structure or functioning of this oxygen carrier molecule was poorly known at the time of Wolverkamp and Waterman's (1960) chapter on respiration in Waterman's "Physiology of Crustacea" and today, 25 years later, although our data base is much expanded our knowledge is still rudimentary compared with that for say mammalian Hb. In functional terms we have abundant evidence that the crustacean hemocyanins (Hcy) are, in fact efficient oxygen carrier molecules, usually of moderate oxygen affinity, higher than usual cooperativity but lower than usual oxygen carrier capacity. Other important roles in carbon dioxide transport, in regulation of hemolymph acid-base balance, and as hemolymph osmo-effector molecules, are now also known but have been very much less well studied.

Since several other papers in this symposium will discuss either the structure of Hcy or functioning of Hcy from other animal groups, this study will thus concentrate solely on the functioning of crustacean hemocyanins. The growth of our knowledge in this area stems largely from the pioneering studies of two workers, Charlotte Mangum from Williamsburg and Jean-Paul Truchot from Paris, the impact of whose studies starting in the early 1970's is significant and well documented in several previous reviews of the functioning of crustacean hemocyanin in oxygen transport (Mangum, 1980, 1983b) and in regulation of acid-base status (Truchot, 1981, 1983). These reviews are, in fact, sufficiently thorough that, in preparation of the current paper it seemed difficult to introduce much that was new either in content or concept. However, since my interest in the hemocyanin molecule has been largely restricted to studies of the molecules function in vivo, I intend to take a fresh, highly functional approach. Since several of the reviews quoted have done an excellent job in collating the many previous studies on Hcy function in vitro I shall, as much as possible ignore this aspect and instead concentrate on an admittedly much smaller body of work which has studied Hcy as it functions in vivo. This paper will attempt to demonstrate the functioning of hemocyanin in gas exchange in animals at rest and in activity under normoxic conditions and during and following acclimation to change in several environmental conditions. Throughout I will try to draw attention to some key questions concerning crustacean Hcy function which remain unanswered, and attempt to show where these in vivo studies can shed some new light. Due to constraints of time and space I shall not attempt to discuss Hcy functioning in acid-base balance (recently reviewed by Truchot, 1981, 1983) or a possible role as an osmoeffector molecule (Snyder and Mangum, 1982).

II. FUNCTIONING OF HEMOCYANIN IN OXYGEN TRANSPORT IN VIVO

A. At rest

Although much less well studied in vivo than in vitro this is nonetheless the best worked aspect of Hcy function. Many authors have measured prebranchial (venous, P_vO_2) and postbranchial (arterial, P_aO_2) oxygen tensions in a variety of crustacean types. Although recent studies (see reviews by Mangum, 1980, 1983b; McMahon and Wilkens, 1983) leave no doubt that the pigment is routinely involved in oxygen transport, some confusion exists in many cases, as to the quantitative importance of its role. Some of this confusion may result from the amazing degree of variability which may be found in the data. Measurements of P_aO_2 may vary from < 10 to > 100 torr (conditions under which many Hcy's may be inoperative) either in normoxic animals of different groups or even in the same species or indeed animals under apparently similar conditions. The question which arises is: can the function (presence) of Hcy in crustaceans really be important given this immense varia-bility ? In this section I will use in vivo data sampled from animals in several physiological states in an attempt to explain the source of much of this variability. Hopefully this will allow us to clarify our understanding of the quantitative role of hemocyanin in O_2 uptake and transport.

Re-examination of a series of experiments carried out in this laboratory and reported by McDonald (1977) and McDonald and McMahon (1977, and unpublished) on the Dungeness crab Cancer magister reveals that changes in physiological state resulting from disturbance or activity prior to sampling can be responsible for much variability. Figure 1 for example demonstrates that, following severe disturbance, ventilation can vary 50 fold (800-16 ml kg/min) in animals which remain essentially quiescent throughout. Similar changes could be shown for the portunid crab Callinectes sapidus (Booth 1982) and in fact for many crustacean species. The decrease results partially from decrease in frequency of scaphogna-thite (ventilatory, f_{sc}) pumping but, importantly, also since one (unilateral pumping) or both (pausing) scaphognathites may periodically cease pumping. Recent evidence indicates that while initially such changes in ventilatory performance may be associated with repayment of an oxygen debt or with removal of CO_2 built up in previous periods of stress, the period of hyperventilation far outlasts the condition it is meant to alleviate and probably results from a stress induced increase in levels of neurohormones such as 5HT, dopamine or the longer acting neuropeptides which are known to increase ventilatory and cardiac pumping (McMahon and Wilkens, 1983). Whatever their cause we should not be surprised to see that these enormous changes in ventilatory level, occurring without change in activity, have

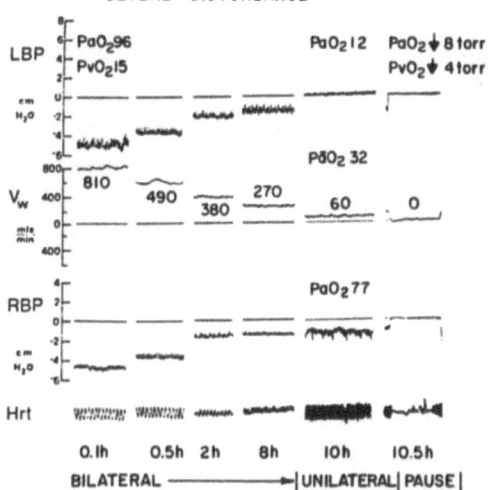

VENTILATORY + CIRCULATORY PERFORMANCE FOLLOWING SEVERE DISTURBANCE

FIG. 1. Changes in circulating hemolymph oxygen tensions associated with variation in ventilatory state without change in activity in the crab Cancer magister. LPB, RBP, left and right branchial pressure traces. Vw, Ventilation volume: flowmeter trace. Hrt, Heart impedance trace. PaO_2, PvO_2, post- and pre-branchial oxygen tensions. Data from McMahon et al., 1979.

dramatic effects on hemolymph oxygenation. These can be demonstrated by hemolymph samples taken from C. magister at each of the 3 levels of excitatory state discussed above (Fig. 2). In this figure measured hemolymph PO_2 levels are plotted against a framework of oxygen equilibrium curves constructed at 8 °C and pH 7.84-8.10 (levels typical of this situation in vivo) using the relationship ($\Delta \log P_{50}/pH = 0.79$; CO_2 buffered) from McDonald (1977). The excited (bilateral pumping) state is clearly associated with very high PO_2 levels. P_aO_2 (96 $\pm$ 12 torr) is far above that which is needed to saturate the hemocyanin, while the circulating P_vO_2 (15 $\pm$ 5 torr) is not low enough to allow much depletion of the hemocyanin bound O_2 reserves. In fact under these conditions less than 50% of the O_2 delivered to tissue is delivered via Hcy. Very similar levels of PO_2 were obtained for C. magister using indwelling cannulae by Johansen et al. (1970).

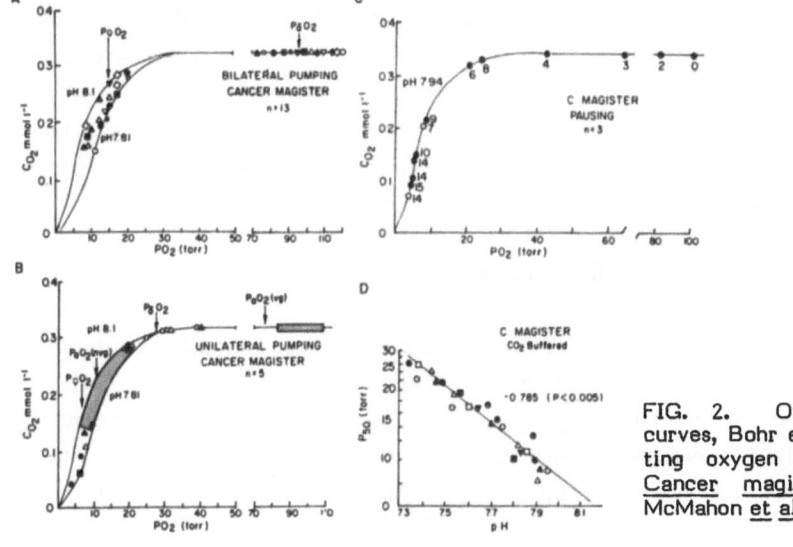

FIG. 2. Oxygen equilibrium curves, Bohr effect and circulating oxygen levels in resting Cancer magister. Data from McMahon et al., 1979.

The picture which emerges from samples taken during unilateral pumping is quite different. Samples taken from the ventilating side are only slightly lower (P_aO_2 = 77 $\pm$ 5 torr) than those reported during bilateral pumping. Samples drawn sequentially from the non-ventilated side however are much lower (P_aO_2 = 11 $\pm$ 7 torr). Mean P_aO_2 determined from the oxygen contents of hemolymph from right and left sides and the oxygen dissociation curves at equivalent pH gives an average P_aO_2 of 28 torr for mixed blood exiting the heart. For comparison mixed samples (R+L) from the ventricule of the heart of a single animal gave values of 81 $\pm$ 4 and 32 $\pm$ 6 torr for bilateral and unilateral pumping respectively (Fig. 2). Under these conditions P_aO_2 and P_vO_2 nicely straddle the steep portion of the O_2 equilibrium curve and Hcy contributes the majority (over 80%) of the oxygen delivered to tissues.

During the least excited state (pausing) animals exhibit periods where the frequency of branchial and cardiac pumping is virtually zero. Thus, except for the very small amount of O_2 trapped in the branchial chamber water, the animal is dependent totally on the venous reserve oxygen bound to Hcy (Fig. 2). Using calculations based on a mean O_2 capacity of 0.61 mM O_2/l, a quiescent $\dot{M}O_2$ of 0.15 mM O_2/kg/min, a typical rate for this species during unilateral pumping and a blood volume of 30%, the venous reserve would actually support the normal rate of $\dot{M}O_2$ for eight minutes: longer in fact, since the $\dot{M}O_2$ of the heart and scaphognathites has been reduced and these may account for a measurable fraction of quiescent $\dot{M}O_2$ (Burggren and McMahon, 1983; Burnett and Bridges, 1981).

It is obvious that in this crab much of the variability seen in PO_2 under experimental conditions stems from changes in excitatory state resulting from experimenter manipulation, disturbance, etc. The situation is compounded by the fact that the animal can, and does change ventilatory state very rapidly. This can obviously cause changes in PO_2 of hemolymph passing through the small gill and pericardial cavity volumes within a matter of seconds. This consideration is of major importance in monitoring levels of P_aO_2 which thus can vary greatly as a result of the sampling procedures. Consequently, extreme caution must be used in the interpretation of values, particularly of P_aO_2, taken from disturbed animals. The use of indwelling catheters or electrodes to measure PO_2 (Angersbach and Decker, 1978; Johansen et al., 1970) may reduce these concerns. However these techniques themselves are not without problems. They are often very short lived and/or usually require that the animals be restrained or tethered. Thus they rarely yield values typical of resting animals as mentioned above.

B. Modification of Hcy function in activity

The importance of the role played by Hcy during activity can be illustrated using the results of several studies carried out over the last 5-10 years in a range of decapod crustacean species (reviewed by McMahon, 1982). McMahon et al. (1979) discussed the mechanisms involved in increased oxygen transport during activity in C. magister but used circulating O_2 tensions typical of bilaterally ventilating animals to characterize the situation at rest. Following the discussion above however I think it more instructive to use values typical of unilateral pumping which we have seen is better representative of resting animals. The dissociation curve in figure 3 thus uses averaged values of P_aO_2 from the ventilated and non-ventilated sides to represent resting P_aO_2 values. In activity the animal

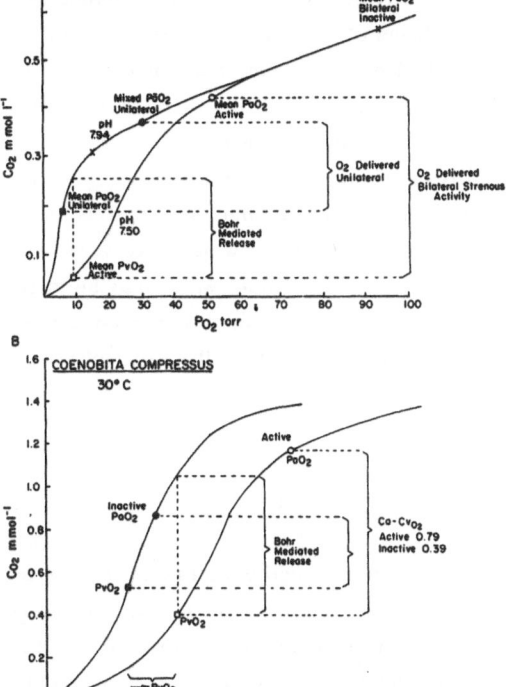

FIG. 3. Oxygen uptake and delivery at rest and in activity in two crabs Cancer magister and Coenobita compressus. Importance of the Bohr effect.

switches to bilateral pumping and thus there is increase in both P_a and P_vO_2. The situation differs from that observed during bilateral pumping in quiescent (disturbed) animals, not only because the increased O_2 demand but also because a substantial Bohr shift results from build up of CO_2 and metabolic (largely lactic) acid which accompanies activity in this species. This has considerable quantitative importance here (see Fig. 3) since marked utilization of the venous reserve O_2 is seen despite an increase in P_vO_2. The Bohr shift thus allows both increase in the gradient between hemolymph and tissues as well as maximizing oxygen unloading to tissues. Both combine to maximize O_2 release during strenuous activity in C. magister. An essentially similar situation occurs following several hours of entirely voluntary activity in the land hermit crab Coenobita compressus (Fig. 3; Wheatly at al., 1985). Due partially perhaps to the voluntary nature of the activity but also to the extremely large O_2 capacity, depletion of the venous reserve is much less pronounced in Coenobita. Comparison of responses to activity both in general and of the O_2 transport system particularly, between the marine C. magister and Callinectes sapidus (Booth and McMahon, 1982) and the land crabs C. compressus (Fig. 3) and Cardisoma carnifex (Wood and

Randall, 1981) suggests that the responses to strenuous activity may be essentially similar in widely divergent crustacean species. The results shown here for C. magister do not really allow a rigid quantitative interpretation since lactate produced during exercice is itself a potent modulator of Hcy affinity in this species (Mangum, 1983c) and would act to reduce the magnitude of the Bohr effect in figure 3a (see later discussion). However the hemocyanin of C. compressus is insensitive to lactate (Wheatly et al., 1985) and thus the essentially similar situation resulting after several hours of voluntary activity seen in this species (Fig. 3b) more accurately displays the quantitative importance of the Bohr effect. It is interesting to note here that the size of the Bohr and lactate effects seem to be quantitatively related at least in decapod crustaceans (Mangum, 1983c).

So to answer the question in vivo data clearly demonstrate that crustacean Hcy works very well facilitating O_2 uptake and delivery both in quiescent animals and in animals showing maximum activity. It is least effective in disturbed animals which have very high ventilatory and cardiac activity but low oxygen consumption, conditions which may be more typical of laboratory experiments than of the natural habitat. The Bohr shift is seen to play a very important role in adjusting O_2 uptake and transport systems to deliver the increased O_2 needed to fuel activity.

III. EFFECTS OF VARIATION IN O_2 CAPACITY

Mangum (1983b) whilst agreeing that the presence of Hcy confers a real advantage to crustaceans repeats an often asked question as to why it is usually present in concentrations rather lower than those of many invert (or vert) Hb's. I would like to re-examine this question in functional terms, i.e. by examination of in vivo data, in order to ascertain what the effects of a reduced concentration of Hcy are, and whether the natural concentrations of Hcy are really limiting.

McDonald (1977) examined the first sub-question directly, looking at aspects of ventilatory and circulatory performance in a series of individual C. magister exhibiting (naturally) a 5 fold range of Hcy concentrations (0.133-0.602 vol% C_{Hcy}^{max} O_2). Theoretically Hcy acts not only to increase the oxygen carrying capacity of the blood but also acts to facilitate diffusion of oxygen across the exchange surfaces at the gills and tissues. Thus, if Hcy content is reduced, we could expect to see compensatory responses in both oxygen uptake and oxygen transport systems. Several responses can be envisaged. The first, that oxygen comsumption (quiescent, routine) declines as a function of Hcy concentration does not seem to occur, at least in C. magister. Compensatory mechanisms thus must occur elsewhere in the system. Spoek (1974) measured a 50% increase in respiratory pumping rate in lobsters experimentally bled to induce 50% decrease in Hcy and assumed that compensatory increase in ventilatory pumping occurred. However he also observed an increase in oxygen consumption and thus the additional ventilation was perhaps more likely associated with disturbance (see above) than with the anemia per se. Ventilation volume is not significantly correlated with hemolymph oxygen capacity in C. magister, but there is a significant correlation with the amount of oxygen extracted from the ventilatory water stream (Fig. 4) i.e. an increase in the efficiency of oxygen uptake across the gills.

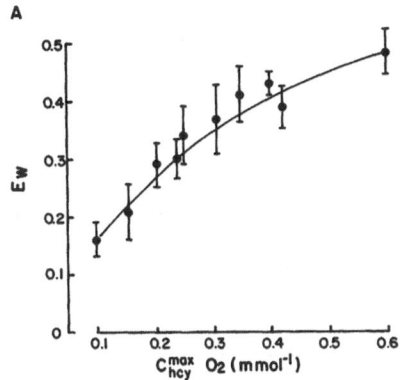

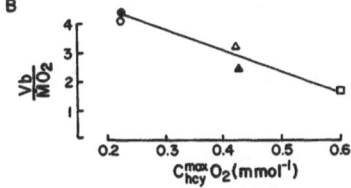

FIG. 4. Effect of reduction in Hcy content (expressed as change in O_2 capacity). A: on efficiency of extraction of O_2 from the branchial water. B: on hemolymph convection requirement. Data from McDonald, 1977.

Several possible mechanisms could be involved. On theoretical grounds we could assume that P_vO_2 might decrease in order to increase PO_2 gradient across the gill surface and/or that P_aO_2 might increase so as to increase the amount of O_2 carried dissolved in the blood, i.e. to compensate for the decrease in O_2 carrying capacity. These changes can be illustrated if we interpolate a relevant resting value of $\dot{M}O_2$ (0.015 mM kg/min) on equilibrium curves constructed for C. magister hemolymph at the two extremes of Hcy content (Fig. 5). While it seems logical that such changes in circulating PO_2 might occur, no significant correlation between either P_aO_2 or P_vO_2 and O_2 capacity could be found in vivo in C. magister, possibly because of the extreme variability discussed above. Nonetheless a re-examination of figure 3b demonstrates that this is exactly what happens in the Coenobitidae which have a much larger O_2 capacity (1.8 mM O_2/liter; McMahon and Burggren, 1979). However, in C. magister, in the absence of changes elsewhere, we must look to the circulatory system and in fact McDonald (1977) was able to demonstrate a significant correlation between cardiac output (Vb) and oxygen capacity (Fig. 4). Although

42

Hcy is clearly involved in oxygen transport, in C. magister at least, the major compensation for anemia apparently occurs in the rate of perfusion both of the gills and tissues. At first glance this may seem to be an energy wasteful response but in fact the response involves change in stroke volume rather than rate (Jorgensen et al., 1982) and may also involve reduction of resistance to decrease the energy required.

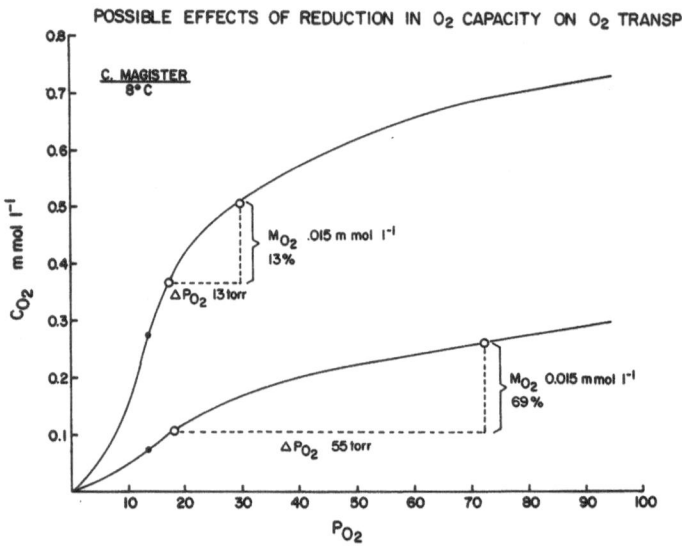

FIG. 5. Theoretical effects of reduction in Hcy capacity on circulating oxygen tensions.

The question as to whether the naturally occurring concentrations of Hcy are limiting has been raised by Mangum (1983b) who comments that the presence of hemocyanin only increases the oxygen carrying capacity of blood by a factor of from 2-5. Her comparison was made at air saturation conditions under which this generalization is often, but not always, true. A more effective comparison, however, can be made if we use more representative O_2 values of in vivo conditions (Table 1). Using these criteria many crustaceans in fact show very much larger amplification of the O_2 carrying capacity than can be seen at air saturation. This is true either of quiescent (undisturbed) animals or of animals during exercice or other conditions where circulating oxygen levels are reduced and the participation of Hcy optimized. We see here that animals from FW (possibly hypoxic) environments and terrestrial animals tend to have both lower circulating O_2 tensions and higher O_2 capacities.

A second factor which should be considered here is that comparisons of oxygen capacity between animals are usually made on the basis of an equivalent blood volume. However, the hemolymph volume of the crustacean open circulatory system is much greater than that of the closed circulatory systems with which comparison is usually made. This tends to exaggerate the difference in overall O_2 capacity. An illustration is provided by

TABLE 1. Factorial Increase in Oxygen Capacity Provided by Hcy for a Range of Crustaceans Compared at Air Saturation (in vitro) and at in vivo Oxygen Levels.

Species		+°	$C^{max}_{O_2}$ Hcy	$C^{max}_{O_2}$ Diss	Fact. Inc.	Pa_{O_2}	Pv_{O_2}	C_{O_2} Hcy	C_{O_2} Diss	Fact. Inc.	State	Reference
			IN VITRO			IN VIVO						
Cancer magister	SW	9°	0.31	0.29	2.1	96 ±-	15	0.27	0.152	2.8	bilateral pumping	McDonald & McMahon (unpub)
			0.31	0.29	2.1	32 ±- 6	7	0.22	0.047	5.6	unilateral pumping	
Callinectes sapidus	SW	20°	0.46	0.22	3.1	79	8	0.33	0.103	4.2	quiescent	Booth (1982)
Cancer productus	SW	10°	0.33	0.27	2.2	51	13	0.22	0.050	5.4	quiescent	DeFur (1980)
Menippe mercenaria	SW	25°	0.62	0.21	4.0	68	6.5	0.35	0.084	5.2	quiescent	Mangum & Mauro (1982)
Homarus vulgaris	SW	15°	0.43	0.24	2.8	49	15	0.23	0.054	5.2	quiescent	Butler et al. (1978)
Pacifastacus leniusculus	FW	13°	0.51	0.28	2.8	34	17	0.15	0.032	5.7	quiescent	Wheatly & McMahon (1982)
		20°	0.44	0.28	2.6	9	5	0.13	0.007	25	quiescent	Rutledge (1981)
Orconectes rusticus	FW	15°	0.51	0.28	2.8	20	4	0.04	0.030	13.5	quiescent	Wilkes & McMahon (1982)
Procambarus clarkii	FW	15°	0.42	0.28	2.5	9	6	0.13	0.005	23	quiescent	Stuart (1984 unpub)
Birgus latro	T	25°	0.81	0.22	4.7	27	13	0.40	0.021	20	quiescent	McMahon & Burggren (unpub)
Coenobita clypeatus	T	23°	1.61	0.22	8.3	16	11	0.89	0.007	128	quiescent	McMahon & Burggren (1979)

comparison of similar sized (300 g) crab (Cancer productus, deFur, 1980) and flounder (Platichthys stellatus, Wood et al., 1979) from the same benthic environment (Table 2). With an O_2 capacity of 6 vols% and a circulating blood volume of 6% body mass the fish has a total O_2 content of approx. 1.1 ml O_2. The crustacean has an O_2 capacity of only 1.25 vols% but a blood volume of 30%, and thus has a total O_2 capacity of 1.25 ml i.e. slightly greater than that of the fish. A possible problem here, that the larger volume of hemolymph may require more effort to circulate is, at least in the comparison above, of reduced consequence due to the lower resistance to flow in the crustacean open circulatory system (Jorgensen et al., 1982).

TABLE 2. O_2 Storage Capacities

	CO	System	Carrier	% Body Volume	ml.O_2 Stored
Platichthys (flounder)	6 vols %	Closed	Hb	6 %	1.1
Cancer (crab)	1.3 vols %	Open	Hcy	30 %	1.3

The conclusion reached from this data is that the problems of low O_2 capacity in crustaceans may have been somewhat exaggerated in the literature since neither the effects of the high blood volume nor low O_2 circulating tensions are usually taken into account. In terms of "normal levels" of Hcy as a limiting factor crustaceans can maintain resting O_2 consumption even at substantially reduced Hcy levels. The fueling of increased tissue oxygen demand as occurs in activity (see mechanisms above) or with temperature increase (see mechanisms below) however, may be curtailed severely and at this level reduced Hcy concentrations may be limiting. It is interesting in this context that molting animals which have very low levels of Hcy also show substantially reduced activity. Many factors, in addition to the performance of the respiratory carrier, can, however, be implicated in limitation of gas exchange (Piiper and Scheid, 1972). At the present time there is no evidence for a conclusion that the naturally occurring concentrations of Hcy are a particular limitation on gas exchange.

IV. ADAPTATION AND ACCLIMATION OF O_2 BINDING

A further question which has intrigued several previous reviewers is the extent to which the O_2 binding characteristics of Hcy can be altered so as to allow efficient oxygen uptake and transport over the full range of conditions to which the animals containing it may be exposed. Actually this breaks down into two related questions:

1. Is Hcy able to change to meet the needs of an animal experiencing a change in environmental conditions ? I will refer to these phenotypic changes which occur in the animal's life time as acclimation.

2. Is Hcy malleable genetically i.e., do animals from different environments have genotypically different Hcy's ? I will refer to these changes in the animals genotype as adaptation.

I would like to start with the animals ability to respond phenotypically to environmental change firstly looking at acute and then long term or seasonal responses. Changes in Hcy O_2 binding characteristics occurring in response to change environmental temperature, oxygen and salinity have been well studied. Due to constraints of space and time I will not be able to deal extensively with changes in response to salinity change. These have been reviewed recently by Mangum (1983a, 1983b).

A. Hypoxia

Compensatory adjustments to Hcy performance in response to hypoxic exposure, both in the lobster Homarus vulgaris (gammarus) (Butler et al., 1978) or crayfish Orconectes rusticus (Wilkes and McMahon, 1982a) include an initial, hyperventilation induced hypercapnia which decreases H^+ and thus results in a negative normal Bohr shift which allows some enhanced O_2 uptake from the depleted water (Fig. 6). However as hypoxic exposure (above Pcrit) continues, ventilation falls, PCO_2 rises, yet O_2 affinity remains high. Clearly a second modulator is active here. Supporting evidence is seen if we plot in vivo values for oxygen partial pressure and content sampled under both normoxic and hypoxic conditions on oxygen equilibrium curves for O. rusticus (data from Wilkes and McMahon, 1982a). The

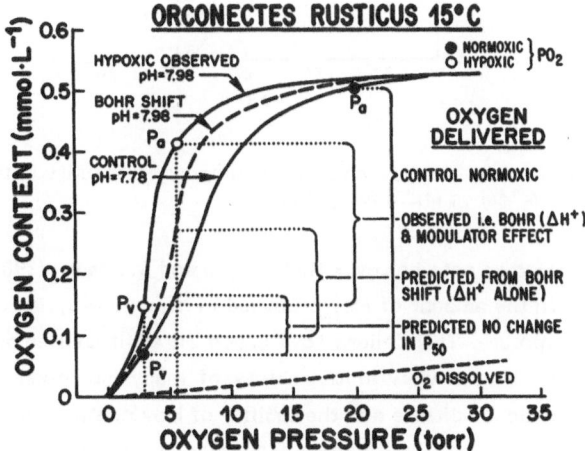

FIG. 6. Compensation for hypoxic exposure in the crayfish Orconectes rusticus: changes in circulating oxygen levels, oxygen affinity and delivery of oxygen to tissues. Data from Wilkes and McMahon, 1982.

resulting plot (Fig. 6) demonstrates a marked additional effect over and above the Bohr shift. Wilkes and McMahon (1982b) have shown that in O. rusticus 23 days following exposure to hypoxia, oxygen affinity is higher at any pH confirming the presence of some other modulating factor. Wilkes and McMahon (1982b) were not able to demonstrate significant changes in the ionic composition of chronic hypoxic exposed O. rusticus (although the concentration of Ca^{2+} increased) and thus were unable to identify the factor. Lactate was not measured. Potent effects of lactate in reduction of O_2 affinity have been demonstrated (Truchot, 1980) and see figure 7. Since hemolymph lactate levels increase markedly under hypoxic exposure in many other crustaceans including crayfish (Wheatly and Taylor, 1981) it seems at least possible that lactate and/or Ca^{2+} are the missing modulators here.

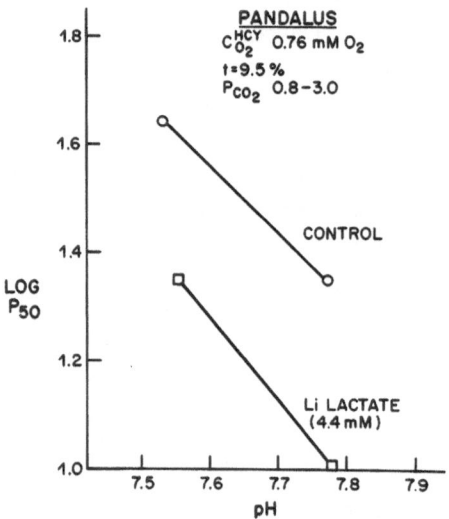

FIG. 7. Potentiating effects of lactate on hemocyanin oxygen affinity in the prawn Pandalus platyceros. Data from McMahon and Booth, 1983.

It is also interesting to note however that Wilkes and McMahon (1982b) were unable to demonstrate increase in the amount of Hcy produced in hypoxic crayfish even after 3 weeks maintained hypoxic exposure. This failure to produce an additional amount of hemocyanin during hypoxia is interesting in view of the ability of many Hb producing crustaceans to increase Hb under hypoxic conditions and the lability of Hcy content in crustaceans exposed to other environmental stresses. Pacifastacus leniusculus breaks down Hcy in concentrated media (Wheatly and McMahon, 1982) while Carcinus maenas produces more Hcy in response to dilute media (Gilles, 1977). Clearly the factors controlling Hcy synthesis are complex, involving factors other than tissue oxygenation.

B. Temperature

Effect of temperature change in Hcy function can be seen at three levels; responses to sudden temperature shifts occurring within minutes or hours, seasonal acclimation, and adaptative effects occurring at the level of the genotype. Immediate effects of change in temperature on Hcy function (Fig. 8) include both direct effects i.e., the change in specific heat of oxygenation (ΔH) which markdely reduces O_2 affinity, and indirect effects i.e., increase in H^+ (decrease in pH) which causes an additional but often smaller decrease in O_2 affinity via the Bohr effect. These effects are of course cumulative and may cause substantial changes in O_2 affinity. [Changes in other effectors especially blood ion levels (or SID) would be expected but are not known in detail for any species.]

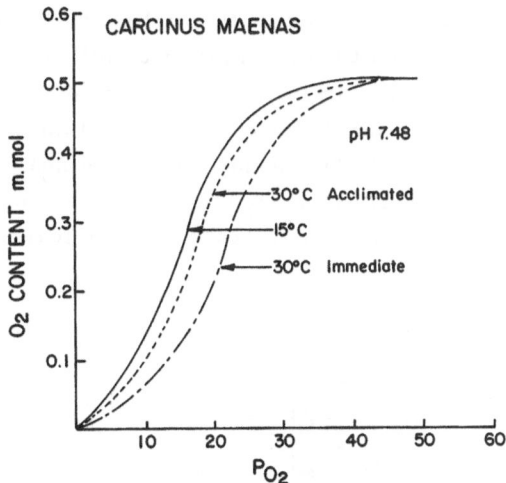

FIG. 8. Seasonal variation in oxygen binding characteristics of <u>Carcinus</u> <u>maenas</u>. Hcy data from Truchot, 1975.

Previous attempts to assess the responses to sudden temperature change have rarely included <u>in vivo</u> measurement of circulating O_2 tensions, but have assumed that the animals show no compensatory changes in PO_2 and thus have attempted to deduce effects of temperature change solely from the changes observed in the oxygen binding characteristics measured <u>in vitro</u>. In fact this assumption is not correct, complex changes in both P_a and P_vO_2 occur as temperature changes in all forms tested. The general pattern that emerges is that as temperature increases so circulating oxygen tensions rise. The increase is most easily demonstrable in venous oxygen tensions where variability is least but is also often apparent at the postbranchial (arterial) level (Fig. 8). At this (arterial) level problems in interpretation arise, especially at temperatures approaching the animals limits when lack of effective response in many systems may combine to limit oxygen uptake. Additionally the variability which we discussed above plays a role and increases where seen, are rarely significant.

The cumulative direct and indirect effects of temperature increase on O_2 binding characteristics of Hcy (data from Booth, 1982; Mauro and Mangum, 1982; McMahon et al., temperature rises is associated with a decrease rather than the expected increase in circulating O_2 contents (Fig. 8). The resultant of these changes is that, as temperature increases, the point representing venous blood moves further down the y axis but up the x axis of the oxygen equilibrium curve (Fig. 8). This allows greater depletion of Hcy bound O_2 (i.e., depletion of the venous reserve) but with increase rather than decrease of oxygen tension. Greater amounts of O_2 thus can be released and an enhanced gradient produced to drive the additional O_2 into the tissues to meet the increased oxygen demand. The converse situation, that as O_2 demand decreases (with decrease of temperature) oxygen carriage is depressed, is also environmentally relevant. In this case, despite the reduction in circulating O_2 tensions a situation nonetheless is eventually reached where at very low temperatures the carrier is very poorly utilized, e.g. at 5 °C in <u>Callinectes sapidus</u> almost all O_2 is supplied from O_2 in solution (Mauro and Mangum, 1982). (We should note here that this dissolved fraction is much increased at these lower temperatures). Poor utilization of the respiratory carrier at low temperatures is not an uncommon finding in animals which become torpid at low temperature and thus show substantially reduced O_2 demand. Although the changes resulting from change in ΔH are often larger, these responses confirm the important contribution made by the Bohr shift in adjusting O_2 transport to meet the minute by minute dictates of tissue metabolism. Increased release of O_2 (participation of Hcy) accompanies increased $\dot{V}_w$ and $\dot{V}_b$ in facilitating O_2 delivery to tissues as temperature rises in crustaceans.

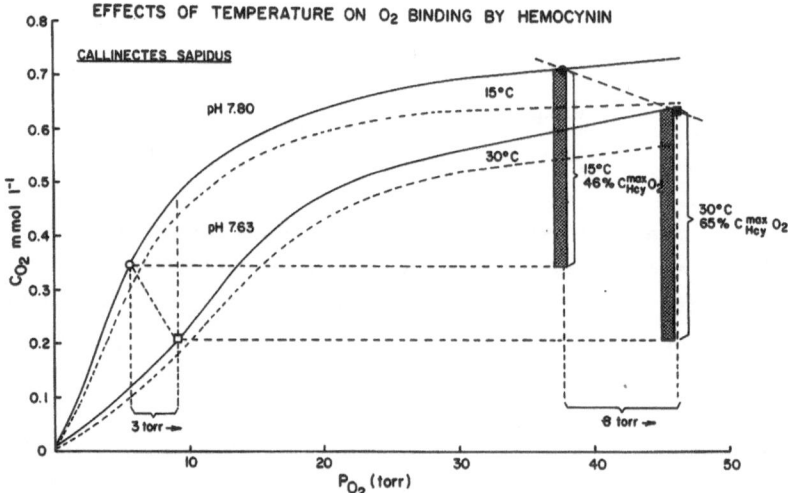

FIG. 9. Variation in circulating oxygen levels and oxygen affinity as temperature increases in the Blue Crab <u>Callinectes</u> <u>sapidus</u>. Broken lines show Hcy bound O_2 alone. Solid lines include dissolved O_2 fraction.

A second important aspect is that of longer term seasonal changes in oxygen affinity such as have been observed for the crabs <u>Carcinus maenas</u> (Truchot, 1975), <u>Callinectes sapidus</u> (Mauro and Mangum, 1982) and the crayfish <u>Pacifastacus leniusculus</u> (Rutlege, 1981) and the prawn <u>Palaemon elegans</u> (Morris <u>et al.</u>, 1984). In these examples the acclimatory effect is often to readjust the oxygen binding characteristics back towards a mean level (Fig. 8). This presumably occurs as other metabolic adjustments at the cellular level occur to allow metabolism to proceed more efficiently at the new acclimation temperature. Similar long-term acclimative changes are seen in portunid crabs in response to salinity change (Mangum, 1983a, 1983b). Crabs acclimated to a low saline estuarine environment have a lower O_2 affinity than crabs acclimated to seawater. Transferred populations show acclimation to readjust O_2 affinity which is more or less complete within 8 days depending on the direction of transfer. Although I do not have any in vivo data to compare with these in vitro changes, I would like to briefly discuss how these longer term changes in Hcy O_2 affinity might be brought about.

Several mechanisms may occur. These include:

Change in level of existing (known) modulators either ionic or organic. Changes in levels of circulating Ca^{2+}, Mg^{2+}, and lactate are observed in hemolymph from Palaemon elegans sampled in summer and winter.

Change in the level of new modulator systems currently suspected but not identified.

Change in aggregate composition of Hcy, i.e. change in Hexamer/Dodecamer or other subunit ratio.

Change in the subunit composition of existing Hcy, i.e. by rearrangement of existing subunits.

Change in the subunit composition of Hcy by production of new (different) subunits, i.e. production of a new Hcy.

Discussion of protein structure is not within the mandate of this article and in any case there are no answers as to which mechanisms are involved but I would like briefly to outline some exciting, if tantalizing results obtained recently (Mangum, 1983a) in this area. From the salinity work discussed above there is some evidence that changes in subunit composition of the Hcy molecule can be correlated with changes in affinity resulting from acclimation to different salinities. Secondly, and this from the season acclimation effects noted above (Mauro and Mangum, 1982), Mangum (1983b) correlates two sets of observations, that of Zatta (1981) showing that the oxygen affinity of Carcinus Hcy is affected by changes in the lipid portion of the molecule and that of Kerr (1969) that seasonal changes occur in hemolymph phospholipid levels. These observations are not yet put together experimentally but the suggestion that O_2 affinity of Hcy could be modulated by change in bound lipid is an exciting possibility well worthy of further study. This is especially true when one considers the number of papers which implicate bound (non-dialysable) co-factors in long term and seasonal acclimation of Hcy binding (Mangum and Mauro, 1982; Morris et al., 1984; Rutledge, 1981; Truchot et al., 1975).

I would like to conclude this section with consideration of adaptative changes. Previous reviews (Mangum, 1980, 1983a) compared the O_2 binding characteristics of a large number of crustacean hemocyanins adjusted to equivalent conditions of temperature and pH and concluded that this molecule was rather conservative, showing very little in the way of change in basic oxygenation properties with habitat, latitude, activity level, etc. At constant pH of 7.5 Mangum showed, however, that crabs from warm environments tested at 25 °C actually had very similar levels of P_{50} to cold water crabs tested at 15 °C suggesting that some adaptation in O_2 binding had occurred in response to the predominant ambient temperature regime. Mangum nonetheless concluded that "change in geographic distribution and or mode of gas exchange must be greater amongst hemocyanin containing crustaceans than amongst other animals to induce an appreciable adaptation in blood oxygen affinity".

The data compiled by Mangum (1983), however, although considerable in number of species, are somewhat limited in that they result from an area of physiological study which has been strongly biased towards reptant marine decapod crustaceans from intertidal and coastal waters. As she notes, considerable exceptions from the "norm" do occur. I gathered together some of these exceptions to see if they had anything in common. In fact two sets of animals did often show oxygen binding characteristics rather different from the "near littoral norm". The first group contained many animals from environments which can become very hypoxic and which had substantially higher oxygen affinities (Fig. 10). This trend is especially notable in FW crayfish many of which seem to have relatively high oxygen affinity but other examples from other groups may occur. A good example is provided by the bathypelagic mysid Gnathophausia ingens whose habitat is the ocean oxygen minimum layer, where ambient oxygen levels rarely rise above 20 torr (Childress, 1971) and which has a particularly high O_2 affinity hemocyanin (P_{50} = 1.5 torr at pH 7.44; Freel, 1978).

The second group contained animals which seemed to have oxygen affinities somewhat less than those of the "normal" range. Included in this group are a variety of open water shrimps, i.e. Pandalus platyceros (McMahon and Booth, 1983) as well as the truly air breathing land crabs such as Birgus latro. (Several "land crabs" such as Coenobita and Cardisoma are actually bimodal or even possibly primarily aquatic breathers (McMahon and Wilkens, 1983) and as such do not have particularly low O_2 affinities). This group can thus be summed up as having a low O_2 affinity blood and living in environments where oxygen lack is relatively rare. We can thus produce a similar sort of range for crustacean Hcy's (Fig. 10) as we are used to seeing in texts for Hb's. This is not to say that all Hcy's will fit into this scheme, exceptions can of course be found, but simply to point out that a fair degree of adaptability does seem to occur in the Hcy molecule at least in response to environmental O_2 availability. The same graph also shows a considerable degree of variability in the magnitude of the Bohr effect. This is apparently very small in those animals which are found in often hypoxic (and presumably hypercapnic) environments where changes in CO_2 and H^+ occur commonly and very large in those animals in which change in H^+ or CO_2 are of small magnitude. The intermediate ranges correspond to the intertidal forms which may experience considerable short term variability.

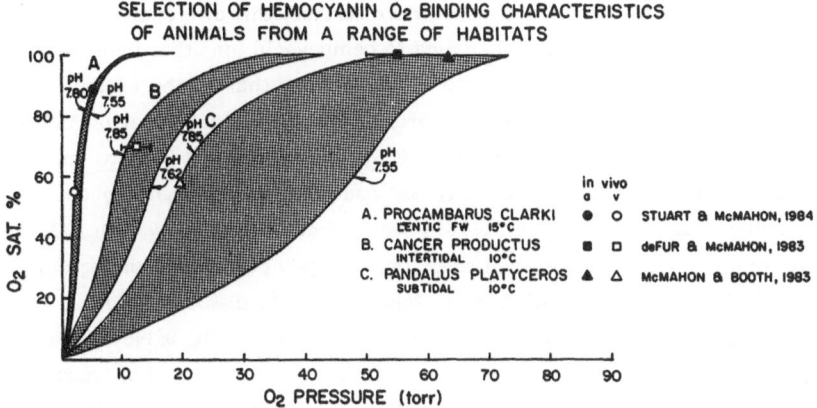

FIG. 10. Adaptation of Hcy function. Hcy O_2 binding characteristics and magnitude of the Bohr effect in Hcy's of animals from different habitats.

Adjustment in Hcy performance to allow effective oxygenation under variable environmental conditions clearly occurs at least three levels. Rapid adjustments may occur directly on the molecule as with temperature effects on Hcy binding or indirectly as in the case of H^+ or lactate ions. In both cases described above the two effects combined to allow better O_2 delivery to tissues in the case of increased temperature to enhance oxygen uptake or as in the case of hypoxia to allow enhanced O_2 uptake. This however is not general, in fact, often the effects of two modulating systems will be opposite leaving oxygen binding essentially unchanged. Mangum and Towle (1977) describe several examples of this latter type which they termed enantiostatic regulation. Maintained or long-term adjustments may involve more stable modulation of the molecule which could involve change in the Hcy structure itself or in a bound component such as phospholipid.

V - THE ROLE OF HEMOCYANIN IN TRANSPORT OF CARBON DIOXIDE: HALDANE EFFECT

Transport of carbon dioxide is as, or more complex than that of oxygen since even smaller amounts (less than 3%) of the CO_2 in crustacean hemolymph occur in simple solution (Fig. 11). The rest is present in bound form either ionic ($HCO_3^- + CO_3^{2-}$) or protein bound as carbamate. Theoretically Hcy would be involved in mobilization from either bound pool and either could be important in CO_2 transport and release. The ionic fraction can be estimated from measurements of CO_2 content, pH, PCO_2, and constants pK'_1 and αCO_2. In all crustaceans studied this is certainly the major fraction but, unfortunately, no direct measurement of the amount of carbamate bound CO_2 occurs for any crustacean species. If total CO_2 content is measured and the contents of HCO_3^-, CO_3^{2-}, and molecular $[CO_2]$ calculated then a discrepancy between the calculated and measured values could be considered equivalent to carbamino bound CO_2. For samples of hemolymph taken under resting conditions from most crustaceans we have studied, this discrepancy is actually very low, suggesting that the amount of CO_2 bound to hemocyanin (or other hemolymph protein) is relatively small. The following discussion thus assumes that in the examples used here carbamino bound CO_2 release is of minor importance (see additional discussion below).

Examination of CO_2 equilibrium curves (Fig. 11) determined in vitro for an aquatic portunid crab Callinectes sapidus, Booth (1982) and for the air-breathing anomuran crab Coenobita compressus, Wheatly et al. (1985) demonstrate a relatively high CO_2 capacity. This results from the buffering action of substances (largely protein) occurring in crustacean hemolymph which, by reacting with the H^+ formed as CO_2 dissociates allows build up of HCO_3^-. Since most protein in hemolymph (at least during intermoult) is Hcy (Busellen , 1970; Mangum et al., 1984) this buffering of extracellular H^+ is an important function of Hcy which we will not have time to discuss here (Truchot, 1981; 1983 for reviews) but as has been clearly demonstrated for several crabs (Booth, 1982; McDonald, 1977; Truchot, 1976) non-bicarbonate buffering power increases proportionally with O_2 capacity. Buffer value of Hcy as expressed per unit O_2 capacity is, in fact, higher (2.54 meq/mM O_2) than that for human Hb but this is largely a result of the relatively low oxygen carrying capacity of crustacean bloods. As expressed per unit protein concentration the buffering power of Hcy is slightly less than that of human Hb (Truchot, 1976). This means that the pH effect resulting from the Bohr shift is necessarily very small. At the low PCO_2 levels (1-4 torr) characteristic of aquatic animals the capacitance coefficient (amount of CO_2 bound per torr PCO_2) is very high (1.25 mM/torr Callinectes, 1.1 mM/torr Coenobita; see Fig. 11). This indicates that differences in PCO_2 pressure across the gills resulting from CO_2 removal are necessarily small and significant differences difficult to measure experimentally at least under normal quiescent conditions.

Several authors have reported the existence of a Haldane effect (i.e. oxygen mediated CO_2 binding and release) in vitro in the hemocyanin of decapod crustaceans including both reptant (Callinectes, Booth, 1982; Carcinus, Truchot, 1976) natant (shrimp, Morris et al., 1984; Weber and Hagerman, 1981) and "terrestrial" (Cardisoma, Randall and Wood, 1981; Coenobita, Wheatly et al., 1984) decapod hemocyanins. Such effects are presumably widespread if not universal in crustacean Hcy's but again are commonly very small and thus hard to demonstrate. Clearly their magnitude will vary with the amount of Hcy and thus the effects may be more easily demonstrated in the land crabs and in shrimp, which have substantially larger Hcy contents.

Despite the apparently small magnitude of the Haldane effect, Truchot (1976) suggested that Hcy may nonetheless play a significant role in CO_2 transport. Recent studies in this laboratory can be used to attempt to quantify the effect and its significance. Measured in vivo values for pre- and postbranchial hemolymph CO_2 tension (P_aCO_2, P_vCO_2) pre- and postbranchial oxygen content and oxygen capacity, together with CO_2 dissociation curves for oxy- and deoxygenated sera (Fig. 11) are required. Note firstly that $[CO_2]$ values were measured utilizing acid extraction and thus include any carbamino-bound CO_2 and also that these PCO_2 values are measured from in vivo samples and are assumed to be at

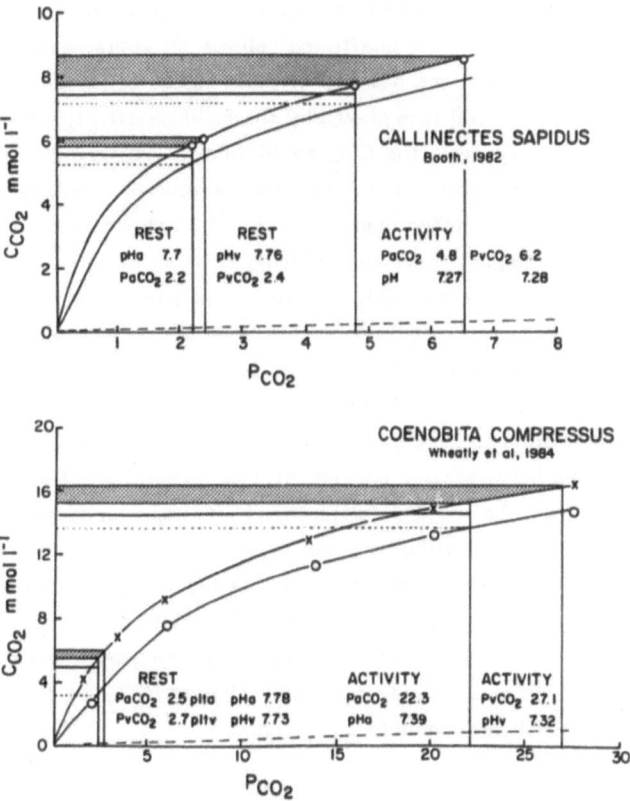

HALDANE EFFECTS : CRUSTACEAN HEMOCYANINS

FIG. 11. CO_2 equilibrium curves and the role of the Haldane effect in CO_2 transport in two crustaceans from different habitats. Broken lines maximal Haldane effect. Solid lines Haldane effect as calculated from in vivo levels.

equilibrium. A computer simulation by Aldridge and Cameron (1979) based largely on assumed levels of CO_2 production and circulating CO_2 tension, suggests that full equilibrium may not be reached in the crab circulation in vivo. However since neither the existence nor magnitude of the disequilibrium is proven and the assumptions have been questioned (Burnett, 1982) the present study has had to assume that for all practical purposes, the PCO_2 levels measured are representative of those which occur in hemolymph approaching and leaving the gill.

CO_2 transport at rest

At rest the PCO_2 differential between pre- and postbranchial hemolymph is small ($P_a - P_v CO_2$ = 0.2 torr Callinectes, 0.3 torr Coenobita) as predicted above. These values are essentially similar to most literature values (reviewed by McMahon and Wilkens, 1983). If we plot these values on the CO_2 equilibrium curve (Fig. 11, deoxy = prebranchial hemolymph) we can, bearing in mind the assumptions above, use the standard textbook approach to assess the amount of CO_2 removal resulting from the Haldane effect. The pre-postbranchial PCO_2

difference is used to estimate the amount of CO_2 which could be released from the gill without intervention of the Haldane effect. We see that, despite the very small a-v PCO_2 difference, the high CO_2 capacitance coefficient allows an appreciable amount of CO_2 (0.2-0.4 mM CO_2/liter blood flow, see hatched areas in figure 11) to be removed at rest. We can now estimate the potential role played by the Haldane effect by moving down the line of constant PCO_2 (Fig. 11) from the CO_2 equilibrium curve for deoxygenated Hcy to that for oxygenated Hcy. The difference between the two curves represents the maximum amount of bound CO_2 (mM/liter; dotted lines in figure 11) which would be mobilized via the Haldane effect assuming complete oxygenation/deoxygenation of Hcy. In Callinectes, for example, this would amount to 0.6 mM CO_2 per liter hemolymph passing through the gills. Of course an oxygen change of this magnitude almost never occurs at rest but an estimate of the magnitude of the Haldane effect in vivo can be obtained from circulating oxygen tensions, contents, hemolymph flow and the oxygen capacity, where these are available. Such estimates for both Callinectes and Coenobita are shown as solid lines in figure 11. Very good agreement is seen between the amount of CO_2 mobilized (7 µM min) and that excreted as calculated from the a-v CO_2 difference in this species (6 µM) also calculated from the data of Booth (1981). Further support is obtained when resting $\dot{M}CO_2$ data are included yielding and RQ of 0.76 a very reasonable resting value for animals on a largely protein diet. These calculations, of course, assume the relationship between oxygen and carbon dioxide binding and release is both equal and linear. Nonetheless a potentially very important role for the Haldane effect in mobilization of CO_2 at the gill surface is demonstrated by analysis of these complete in vivo data.

CO_2 transport in activity

Figure 11 also details P_a and P_vCO_2 from Callinectes and Coenobita hemolymph samples taken following sustained moderate activity. If we assume that the rise in CO_2 production follows that in O_2 consumption (R = 1 approx.) then the animal must remove more than 3 fold (Callinectes) to 6 fold (Coenobita) the CO_2 removed at rest. Hemolymph PCO_2 also rises in activity due to the change in CO_2 equilibrium resulting from a pronounced hemolymph lactacidosis (Booth at al., 1982; Wheatly et al., 1984). Removal of this additional CO_2 seems not to be as easy as we usually assume, since circulating PCO_2 levels rise substantially. The rise in PCO_2 in Callinectes is remarkable since this is an aquatic animal and we have been led to believe that retention of CO_2 is not possible, especially during the marked hyperventilation (Booth et al., 1982) which accompanies activity in this species. The increase in PCO_2 in Coenobita is massive but perhaps easier to explain since this is an animal which has access to water stored in the shell at rest (McMahon and Burggren, 1979) but may lose access to this water, due to postural changes while active. Thus the increase in PCO_2 levels in this case may, in part, be associated with the transition to aerial respiration in addition to exercise. Nonetheless both sets of results suggest strongly that mechanisms for CO_2 removal become swamped in activity.

It is interesting to note that not only do circulating PCO_2 levels increase but that increase also occurs in the a-v PCO_2 difference which increases 8-10 fold in either animal. Examination of the CO_2 dissociation curves may offer some explanation for this. Firstly more CO_2 enters the circulation and thus P_vCO_2 increases. Further increase results from acidosis as described above and the situation may be progressive if limitation occurs on maximal rate of CO_2 release. This increase in circulating CO_2 tensions causes a titration up to the CO_2 equilibrium curve to regions where the capacitance coefficient for CO_2 is reduced and thus where much larger changes in PCO_2 per unit CO_2 loss would be recorded.

Is the Haldane effect of increased value in activity ? The actual magnitude can only be expected to rise as more oxygen is bound or freed from Hcy at the gills/tissues respectively. This increase is relatively small in swimming activity in Callinectes but larger in Coenobita (Fig. 11) where the a-v O_2 content difference is increased substantially (Fig. 3). In both cases the actual contribution of the Haldane effect to CO_2 excretion is dependent on the rate of gill blood flow and thus amplified by the increase in Vb. resulting during activity in both species (Booth et al., 1982; Wheatly et al., 1985).

Despite considerable differences in structure and habitat there is apparently little difference in the process of CO_2 transport between the two species. There is clear evidence for marked limitation to CO_2 excretion during activity in both species but particularly in Coenibota. Interestingly normal CO_2 levels are reached very quickly in this species on return to resting situations where water access can occur. This suggests that the gills play a very important role in CO_2 excretion in Coenobita. This is perhaps due to the presence of the enzyme carbonic anhydrase which is found in high concentration in the gills and which may play a major role in CO_2 excretion in these crabs during activity.

It is clearly apparent that in addition to the role played in transport of O_2 Hcy also plays an important role in CO_2 transport. As with any other respiratory pigment studied competition occurring between the two functions is crucial to the efficiency of gas transport and exchange. Just as earlier we saw that the Bohr effect was very important in adjusting the amount of O_2 delivery to tissues during activity etc. so later we have seen that the Haldane effect may play an equally important role in transport and removal of CO_2. The mechanisms involved in these effects are not as well understood for Hcy as for Hb. Especially the role of carbamino bound CO_2 is not known.

There has been a persistent attempts in the literature to downgrade the properties of hemocyanin to a rank below that of Hb as a functional respiratory carrier. I do not think that this is really valid. Hcy seems to work very well for the larger open circulatory systems in which it is usually found. It is, in my opinion, an adaptable molecule and also is influenced by a wide range of modulator systems which can modify its basic properties to suit gas exchange to wide variability in the animals habitat. It does not seem to confer disadvantages on the animals which have it. Indeed in the hemocyanin containing cephalopod molluscs we seem to see the pinnacle of success of the invertebrate phyla.

REFERENCES

Aldridge JB, Cameron JN (1979) CO_2 exchange in the blue crab, <u>Callinectes</u> <u>sapidus</u> (Rathbun). J. Exp. Zool. 207:321-328

Angersbach D, Decker H (1978) Oxygen transport in crayfish blood: effect of thermal acclimation, and short term fluctuations related to ventilation and cardiac performance. J. Comp. Physiol. 123:105-112

Booth CE (1982) Respiratory responses to activity in the blue crab, <u>Callinectes</u> <u>sapidus</u>. Ph.D. Thesis. University of Calgary, Calgary, Alberta, Canada

Booth CE, McMahon BR, Pinder AW (1982) Oxygen uptake and the potentiating effects of increased hemolymph lactate on oxygen transport during exercice in the blue crab, <u>Callinectes</u> <u>sapidus</u>. J. Comp. Physiol. 148: 111-121

Burggren WW, McMahon BR (1983) An analysis of scaphognathite pumping performance in the crayfish <u>Orconectes</u> <u>virilis</u>: compensatory changes to acute and chronic hypoxic exposure. Physiol. Zool. 56: 309-318

Burnett LE (1982) CO_2 excretion across isolated perfused crab gills: facilitation by carbonic anhydrase. Amer. Zool. 24: 253-264

Burnett LE, Bridges CR (1981) The physiological properties and function of ventilatory pauses in the crab <u>Cancer</u> <u>pagurus</u>. J. Comp. Physiol. 145: 81-88

Busselen P. (1970) The electrophoretic heterogeneity of <u>Carcinus</u> <u>maenas</u> hemocyanin. Arch. Biochem. Biophys. 137:415-420

Butler PJ, Taylor EW, McMahon BR (1978) Respiratory and circulatory changes in the lobster (<u>Homarus</u> <u>vulgaris</u>) during long term exposure to moderate hypoxia. J. Exp. Biol. 73: 131-146

Childress JJ (1971) Respiratory adaptations to the oxygen minimum layer in the bathy-pelagic mysid <u>Gnathophausia</u> <u>ingens</u>. Biol. Bull. 141: 109-121

Freel RW (1978) Oxygen affinity of the heomlymph of the mesopelagic mysidacean <u>Gnathophausia</u> <u>ingens</u>. J. Exp. Zool. 204: 267-274

deFur PL (1980) Respiratory consequences of air exposure in <u>Cancer</u> <u>productus</u>, an intertidal crab. Ph.D Thesis. University of Calgary, Calgary, Alberta, Canada

Gilles R. (1977) Effects of osmotic stresses on the protein concentration and pattern of <u>Eriocheir</u> <u>sinensis</u> blood. Comp. Biochem. Physiol. 56A: 109-114

Johansen K, Lenfant C, Mecklenburg TA (1970) Respiration in the crab, <u>Cancer</u> <u>magister</u>. Z. Vergl. Physiol. 70: 1-19.

Jorgensen D, Bourne G, Burnett L, deFur P, McMahon B (1982) Circulatory function during hypoxia in the dungeness crab, <u>Cancer</u> <u>magister</u>. Amer. Zool. 22:958A

Kerr MS (1969) The hemolymph proteins of the blue crab, <u>Callinectes</u> <u>sapidus</u>. II. A lipoprotein serologically identical to oocyte lipovitellin. Dev. Biol. 20: 1-17

Mangum CP (1980) Respiratory function of the hemocyanins. Amer. Zool. 20:19-38

Mangum CP (1983a) Adaptability and inadaptability among HCO_2 transport systems: an apparent paradox. Life Chem. Reports 4: 335-352

Mangum CP (1983b) Oxygen transport in the blood. In: Bliss DE and Mantel LH (eds) The Biology of Crustacea. Vol. 5. Academic Press, pp. 373-429

Mangum CP (1983c) On the distribution of lactate sensitivity among the hemocyanins. Mar. Biol. Letters 4: 139-149.

Mangum CP, Towle DW (1977) Physiological adaptation to unstable environments. Am. Sci. 65: 67-75

Mangum CP, McMahon BR, deFur PL, Wheatly MG (1984) Gas exchange, acid-base balance, and the oxygen supply to the tissues during a molt of the blue crab Callinectes sapidus. J. Crust. Biol. (In press)

Mauro NA, Mangum CP (1982) The role of the blood in the temperature dependence of oxidative metabolism in decapod crustaceans. I. Intraspecific responses to seasonal differences in temperature. J. Exp. Zool. 219: 189-198

McDonald DG (1977) Respiratory physiology of the crab Cancer magister. Ph.D Thesis, University of Calgary, Calgary, Alberta, Canada

McDonald DG, McMahon BR, Wood CM (1977) Patterns of heart and scaphognathite activity in the crab Cancer magister. J. Exp. Zool. 202: 33-44

McMahon BR (1981) Oxygen uptake and acid-base balance during activity in decapod crustaceans. In: Herreid CF and Fourtner CR (eds) Locomotion and Energetics in Arthropods. Plenum Publishing Corp. pp. 299-355

McMahon BR, Burggren WW (1979) Respiration and adaptation to the terrestrial habitat in the land hermit crab Coenobita clypeatus. J. Exp. Biol. 79:265-281

McMahon BR, Burggren WW (1980) Oxygen uptake and transport in three air breathing crabs. The Physiologist 23 (4): 928A

McMahon BR, Wilkes PRH (1983) Emergence responses and aerial ventilation in normoxic and hypoxic crayfish Orconectes rusticus. Physiol. Zool. 56: 133-144

McMahon BR, Wilkens JL (1983) Ventilation, perfusion and oxygen uptake. In: Mantel L and Bliss DE (eds) The Biology of Crustacea, Vol. 5: Internal Anatomy and Physiological Regulation. Academic Press, pp. 289-372

McMahon BR, Sinclair F, Hassall CD, deFur PL, Wilkes PRH (1978) Ventilation and control of acid-base status during temperature acclimation in the crab Cancer magister. J. Comp. Physiol. 128: 109-116

McMahon BR, McDonald DG, Wood CM (1979) Ventilation oxygen uptake and haemolymph oxygen transport following enforced exhaustive activity in the Dungeness crab, Cancer magister. J. Comp. Physiol. 128B: 109-116

Morris S, Taylor AC, Bridges CR, Grieshaber M (1984) Respiratory properties of the haemolymph of the intertidal prawn Palaemon elegans Rathke. J. Exp. Zool. (In press)

Piiper J, Scheid P (1972) Maximum gas transfer efficacy of models for fish gills, avian lungs and mammalian lungs. Resp. Physiol. 14: 115-124

Randall DJ, Wood CM (1981) Carbon dioxide excretion in the land crab Cardisoma carnifex. J. Exp. Zool. 218: 37-44

Rutledge PS (1981) Circulation and oxygen transport during activity in the crayfish, Pacifastacus leniusculus. Am. J. Physiol. 240: R-99 - RO105

Snyder GK, Mangum CP (1982) The relationship between the capacity for oxygen transport, size, shape and aggregation state of an extracellular oxygen carrier. In: Joseph and Celia Bonaventura, and Shirley Tesh (eds) Physiology and Biochemistry of Horseshoe Crabs. A.R. Liss, New York.

Spoek GL (1974) The relationship between blood haemocyanin level, oxygen uptake and the heart-beat and scaphognathite-beat frequencies in the lobster Homarus gammarus. Neth. J. Sea Res. 8: 1-26

Truchot JP (1975) Factors controlling the in vitro and in vivo oxygen affinity of the hemocyanin in the crab Carcinus maenas (L.) Respir. Physiol. 24: 173-189

Truchot JP (1976) Carbon dioxide combining properties of the blood of the shore crab Carcinus maenas (L.): CO_2 dissociation curves and Haldane effect. Comp. Physiol. 112: 283-293

Truchot JP (1980) Lactate increases the oxygen affinity of crab hemocyanin. J. Exp. Zool. 214: 205-208

Truchot JP (1981) L'équilibre acido-basique extracellulaire et sa régulation dans les divers groupes animaux. J. Physiol. (Paris) 77: 529-580

Truchot JP (1983) Regulation of Acid-Base Balance. In: Mantel LH and Bliss DE (eds) The Biology of Crustacea, Vol. 5: Internal Anatomy and Physiological Regulation. Academic Press. pp. 431-456

Weber RE, Hagerman L (1981) Oxygen and carbon dioxide transporting qualities of hemocyanin in the hemolymph of a natant decapod Palaemon adspersus. J. Comp. Phsyiol. 145: 21-27

Wheatly MG, McMahon BR (1982) Responses to hypersaline exposure in the euryhaline crayfish Pacifastacus leniusculus. II. Modulation of haemocyanin oxygen binding in vitro and in vivo. J. Exp. Biol. 99: 447-467

Wheatly MG, Taylor EW (1981) The effect of progressive hypoxia on heart rate, ventilation, respiratory gas exchange and acid-base status in the crayfish Austropotamobius pallipes. J. Exp. Biol. 92: 125-141

Wheatly MG, McMahon BR, Burggren WW, Pinder A (1985) Hemolymph acid-base and blood gas status during sustained voluntary activity in the land hermit crab Coenobita compressus. H. Milne Edwards. Submitted to Journal of Experimental Biology, 1984

Wilkes PRH, McMahon BR (1982a) Effect of maintained hypoxic exposure on the crayfish Orconectes rusticus. II Ventilatory, acid-base and cardiovascular adjustments. J. Exp. Biol. 98: 119-137

Wilkes PRH, McMahon BR (1982b) Effect of maintained hypoxic exposure on the crayfish Orconectes rusticus. II. Modulation of haemocyanin oxygen affinity. J. Exp. Biol. 98: 139-149

Wolverkamp HP, Waterman TH (1960) Respiration. In: Waterman TH (ed) The Physiology of crustacea. Vol. 1. Academic Press. pp. 35-100

Wood CM, McMahon BR, McDonald DG (1979) Respiratory gas exchange in the resting starry flounder Platichthys stellatus: Comparison with other teleosts. J. Exp. Biol. 78: 167-179

Wood CM, Randall DJ (1981) Haemolymph gas transport, acid-base regulation, and anaerobic metabolism during exercise in the land crab Cardisoma carnifex. J. Exp. Zool. 218: 23-35

Zatta P (1981) Protein-lipid interactions in Carcinus maenas hemocyanin. Comp. Biochem. Physiol. 69B: 731-735

Primary Structure of Arthropod Hemocyanins

B. LINZEN, W. SCHARTAU, H.-J. SCHNEIDER

The relatively low solubility of oxygen in water has led, in the animal kingdom, to the development of oxygen carriers which increase the O_2 capacity of the blood between one and two orders of magnitude and thus provide the basis for continuous high activity. There are three types of such carriers which bind oxygen by different principles: (i) The hemoglobins, in which one Fe(II) is bound by a protoporphyrin plus one histidine residue of the globin. This has proved to be the most successful invention which occurs throughout the animal kingdom. (ii) The hemerythrins where two Fe(II) atoms bind one O_2, being themselves complexes by members of the polypeptide chain; the hemerythrins are restricted to a small number of invertebrate animals. (iii) The hemocyanins, in which the active site is a pair of Cu(I) atoms which are also bind directly to the polypeptide chain; there is no heme. While hemoglobins are either intra- or extracellular, hemerythrins are exclusively intracellular and hemocyanins exclusively extracellular.

The primary structure of human α - and β -hemoglobin was solved in 1961 by Braunitzer and his coworkers, and the first hemerythrin structure in 1968 (Klippenstein et al.). For both also the tertiary and quaternary structures are known. The two types of proteins are entirely unrelated whatever the level of organization one looks at.

In contrast, progress in the hemocyanin field has been slow. This was caused by the very complex architecture of these molecules. Arthropod hemocyanins (A-Hc's) are composed of subunits with a Mr about 75,000/2 Cu, larger than a hemoglobin tetramer. These subunits aggregate into hexamers or multiples thereof, and the biggest A-Hc molecule - from the horseshoe crab - is about the size of a ribosome. Molluscan hemocyanins are even larger: their functional unit is a domain of 50 kDa/2 Cu, and 8 domains are combined in one polypeptide chain (Mr ca. 400 000). The native molecule is composed of 10 or 20 of these 400 kDa subunits (Mr ca. $8\text{-}9 \times 10^6$). There are excellent recent reviews by van Holde and Miller (1982) and Ellerton et al. (1983). The problem was aggravated by the fact that the Hc subunits are not homogenous. In the A-Hc's this was first demonstrated by Busselen (1970) for Carcinus Hc, and by Lamy et al. (1970) for Androctonus Hc. It is now firmly established that A-Hc's are built from different subunits, the number of different subunit types being related to aggregate size (review: Linzen, 1983). This had indeed to be expected for a hierarchy of related, closed structures (Klarman and Daniel, 1981), but in reality more different subunits are usually found than required by theory. In the 37S (24-meric) cheliceratan hemocyanins there are 7-8 immunologically distinct subunits. Extensive work in the laboratories of van Bruggen, Lamy, and Linzen (Lamy et al., 1981; Markl et al., 1981; Sizaret et al., 1982) has led to a comprehensive understanding of the quaternary structures of Androctonus australis and Eurypelma californicum. More recently, also the structure of Limulus polyphemus Hc could be tied in (Lamy et al., 1983). Each subunit occupies a specific position and if one tries to reassemble a Hc from a mixture of isolated subunits, one will obtain negative results if only one subunit is omitted (Markl et al., 1982). This high degree of specificity is also expressed by the immunological behaviour: Native subunits of one species but occupying different positions in the oligomer, do not crossreact, subunits from different species, occupying equivalent positions, do crossreact, even if the time of species divergence was some 400 or 500 millions years ago (Lamy et al., 1979; Lamy et al., 1983; Markl et al., 1984; Kempter et al., submitted for publication).

The first comprehensive characterization of all subunits from an A-Hc was performed with E. californicum Hc (Schneider et al., 1977; Markl et al., 1979). The seven subunits differ by molecular weight, isoelectric points, amino acid composition, and fingerprints after tryptic or chemical digestion, in one word: by their primary structure. Considering results obtained for scorpion (Jollès et al., 1979) and horseshoe crab Hc's (Sullivan et al., 1976; Takagi and Nemoto, 1980) we can safely generalize this conclusion. These findings and the methods to isolate individual subunits in quantity constitute a breakthrough which has opened the path to amino acid sequence work.

The Eurypelma Hc subunits d and e can be isolated most easily in pure state and were therefore chosen for sequence analysis. On professor Braunitzer's advice we started with two subunits in parallel - from his experience with the hemoglobin α and β chains he predicted that we would go along faster, taking advantage of the homologies to be expected. This became in fact true, although it created other problems: a doubling of the number of positions to be determined to about 1250 residues, and laboratory competition. The latter was enhanced by the fact that we had access to an automated sequencer for one chain, while the other sequence was done entirely by manual methods.

Considering the length of these chains we were fortunate to find that limited proteolysis of the undenatured subunits produced only two large fragments in each case. Subunit d was cleaved by trypsin about in the middle of the chain, while subunit e after chymotryptic cleavage gave a 29 kDa and a 42 kDa fragment (later on a third fragment of 13 residues was discovered). These large fragments were quite easily isolated after denaturation and carboxymethylation. The problem of fractionating cyanogen bromide and enzymatic digests was thereby reduced tremendously. The sequencing strategy may be demonstrated for subunit d (Fig. 1).

There are 14 methionine residues, so that 15 cyanogen bromide peptides could be expected. All of them were isolated, and 9 could be sequenced completely. The others were subdigested with Staphylococcus V 8 protease or chymotrypsin. The two large fragments

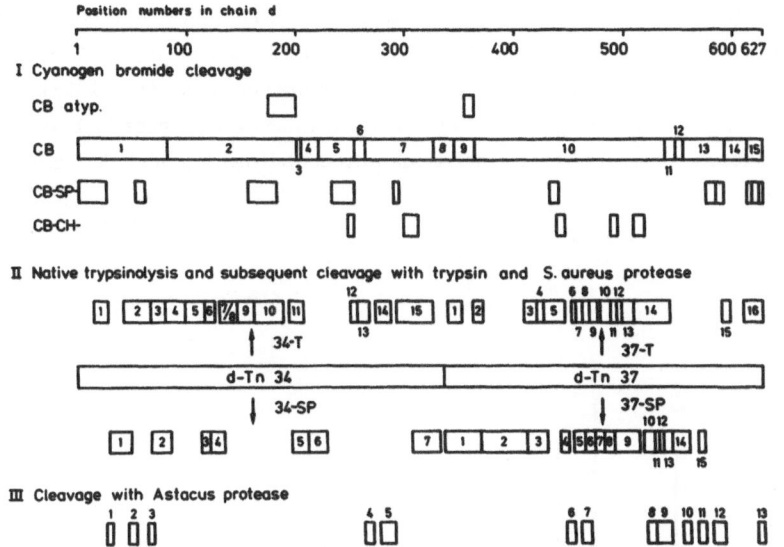

FIG. 1. Sequencing strategy for Eurypelma californicum hemocyanin, subunit d. I. Cyanogen bromide (CB) cleavage produced 15 peptides; "atypical" cleavage occurred at Trp 173 and Asp 354 (acid hydrolysis). CB peptides were sub-cleaved with Staphylococcus protease (SP) or chymotrypsin (CH). II. The native subunit is split into two large fragments by trypsin. d-T$_n$ 34 and d-T$_n$ 37 digested with trypsin (T) or Staphylococcus protease (SP). III. Cleavage with Astacus leptodactylus protease.

after limited proteolysis, d - T$_n$ - 34 (N-terminal) and d - T$_n$ - 37 (C-terminal) were also cleaved with CNBr which allowed to sort the CNBr peptides and to identify CB-8 as the overlap between the two large fragments. The overlaps between the CB-peptides were obtained by cleaving d - T$_n$ - 34 and d - T$_n$ - 37 with trypsin or Staphylococcus protease. Finally, when cleavage positions were not favourable for conventional enzymes, we employed a protease isolated by Zwilling et al. (1981) from the digestive fluid of the crayfish Astacus leptodactylus. This enzyme prefers short nonpolar side chains like Ala, Thr, Gly, Ser.

The peptide mixtures resulting from CNBr or enzymatic cleavage were fractionated first by gel filtration, followed by HPLC. This turned out to be a most powerful method, and by using two different gradient systems on reversed-phase columns, even very complex mixtures could be resolved entirely (Figs. 2 and 3). Manual sequencing was done by the DABITC-method of Chang et al. (1978); since 8 or more peptides can be sequenced in parallel by one person, and two steps performed per day (the record was five steps), the automatic sequencer can be left far behind. We usually worked in the 1-5 nanomole range.

The strategy employed for subunit e was principally similar, except that short peptides were sequenced manually, but the longer ones in the automatic sequencer thanks to collaboration with Drs. A. Henschen and F. Lottspeich. The forecast by Prof. Braunitzer became true - there were so many homologies that a great number of peptides could be placed immediately into the correct position and in turn, the required overlapping peptides pinpointed more easily. Nevertheless, each sequence is completely consistent in itself. The two chains were finished simultaneously in May 1983 (Schneider et al., 1983; Schartau et al., 1983). d contains a total of 627 amino acids with a molecular mass of 72 232, e contains 621 amino acids with a molecular mass of 71 361. Both values are in excellent agreement with values previously determined by analytical ultracentrifugation and SDS-electrophoresis, respectively.

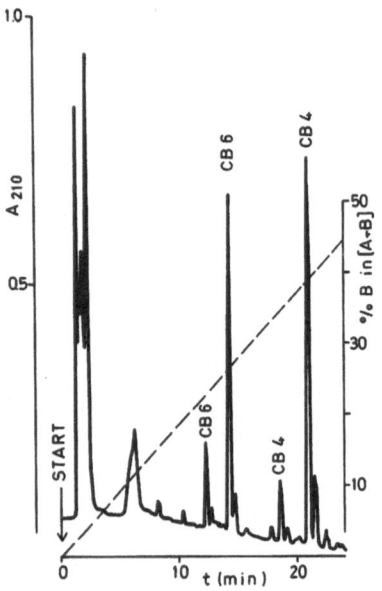

FIG. 2. Separation of CB-peptides 4 and 6 obtained from Eurypelma Hc subunit d, by HPLC. For experimental conditions, see Schartau et al. (1983). Each peptide occurs in two forms, with C-terminal homoserine or homoserine-lactone.

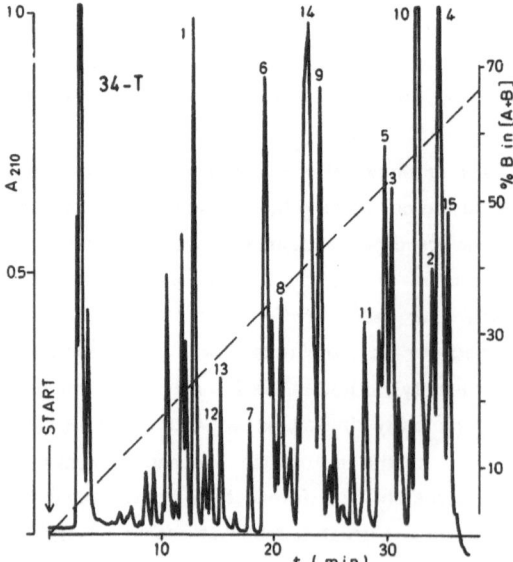

FIG. 3. Separation of tryptic peptides obtained from N-terminal half of subunit d, by HPLC. Numbering is from N to C. For experimental conditions, see Schartau et al., 1983.

In the meantime, the subunit α of Tachypleus tridentatus (a horseshoe crab) Hc has been sequenced completely (Nemoto and Takagi, 1983), and 4 other subunits are nearly complete: Limulus polyphemus Hc, subunit II (A. Riggs and coworkers; cf. Yokota and Riggs, 1984), Eurypelma californicum Hc subunit a (Schartau et al., in preparation), and two crustacean subunits, Panulirus interruptus (a spiny lobster) Hc subunit a (Beintema and coworkers; cf. Gaykema et al., 1984), and Astacus leptodactylus (a crayfish) subunit b (Schneider et al., to be published) (All existing sequence information has been laid down in the Diploma Thesis of W. Voll, Faculty of Biology, University of Munich, 1984, and Thesis submitted in partial fulfilment of State Examination 1985, by R. Grisshammer and by A. Goettgens, University of Munich, 1984.) I shall base all conclusions on the comparison of these 7 completely or partly solved structures, and I am most grateful to all these colleagues for allowing me to quote from unpublished material.

For one thing, the debate on subunit molecular weight can be ended. There are no 50 kDa or 25 kDa subunits nor 'structural' subunits in A-Hc's. On the other hand the relatively high molecular weights determined for crustacean Hc's by SDS electrophoresis - up to 94 kDa - were not confirmed by the results of X-ray crystallography (Gaykema et al., 1984). They must have been caused by an anomalous SDS binding. At present we can state that cheliceratan Hc's are about 625 residues long, and crustacean Hc's about 660 residues.

It was also unexpected that in six of the sequenced chains no carbohydrate was found, Panulirus Hc being the only exception. The carbohydrate attachment site, Asn-Val-Ser-Phe, is not conserved in the otherwise very related crayfish Hc, where the sequence reads Asn-Met-Asp-Phe. Therefore, we cannot attribute yet a physiological significance to the carbohydrate side chain.

In order to align the seven sequences we had to introduce gaps at various points so that the total length was increased to 676 residues. Of these, 17.3% (117 residues) are identical for all chains, 33.6% (212 out of 630 compared residues) are identical for the cheliceratan chains only, and 67.3% (307/465 residues compared) for the crustacean chains. Admittedly, these figures may change somewhat when more sequences can be compared. The degree of homology becomes much more pronounced, if also side chains with similar physical properties are set equal, eg. Ile, Leu, Val, Met, or Phe, Tyr, Trp, or Glu, Asp, or Arg, Lys and His. For all chains, a value of 28.6% homology is attained then, and for all cheliceratan chains 47.3%. Figure 4 ilustrates the degree of homology between all cheliceratan chains. We can go a step further and consider the amino acid side chains in terms of their hydrophylic or hydrophobic behaviour. E.g. in Eurypelma d we find a sequence -Phe-Glu-Glu-Leu-Glu- (starting at position 551), to which corresponds -Val-Lys-Lys-Leu-Arg- in Eurypelma e. Although the charges are opposite, the strongly hydrophilic character is entirely conserved. Hopp and Woods (1981) have generated 'hydrophilicity profiles' by averaging the hydrophilicity values for a group of six amino acids and shifting the group stepwise (by one residue) from the N- to the C-terminus. (The hydrophilicity values are the free energies of transfer from water to organic solvent, e.g. ethanol, or _vice versa_, and are obtained by solubility measurements in mixtures of the organic solvent with water; Nozaki and Tanford, 1971). These profiles are very detailed, almost like a fingerprint. If the cheliceratan subunits Eurypelma d, Eurypelma e, and Tachypleus α are compared, many sections turn out to be almost identical (Fig. 5), and even if such different chains as Panulirus a, Limulus II and Eurypelma a are compared, the similarity of the profiles is striking, in spite of the fact that these hemocyanins must have diverged some 400-600 millions years ago.

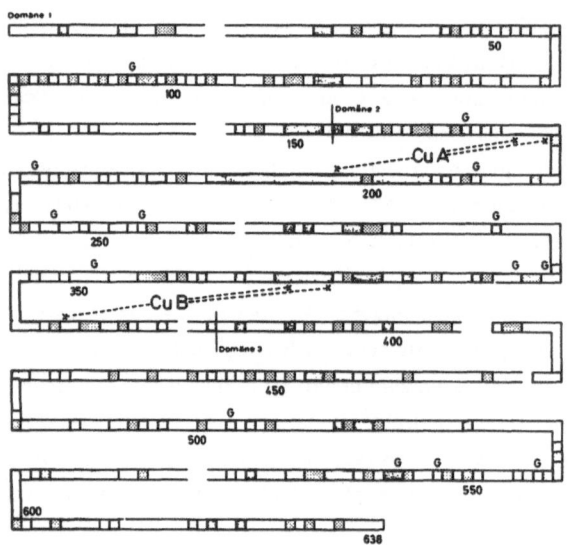

FIG. 4. Homology of amino acid residues for all cheliceratan Hc subunits sequenced so far (Eurypelma californicum a, d, e; Tachypleus tridentatus α ; Limulus polyphemus II). The domains as defined by Gaykema et al. (1984) are indicated. Dark sections: identical residues in all chains; little stippled: residues with similar side chains (isofunctional). G: completely conserved glycine residues. One residue equals one square box. Gaps were introduced, when one of the sequences was not complete yet.

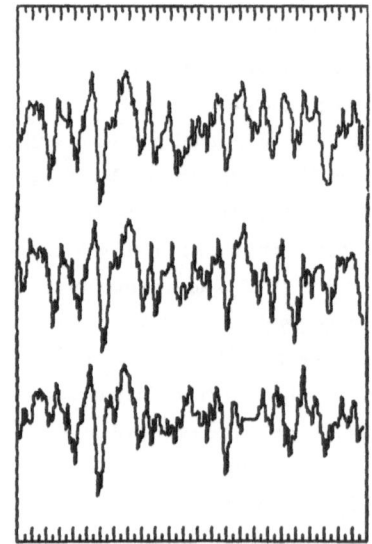

FIG. 5. Hydrophilicity profile (Hopp and Woods, 1981) for the first 250 residues of the cheliceratan Hc subunits Tachypleusα , Eurypelma e, and Eurypelma d. Hydrophilic peaks pointing upward. See text for generation of profiles. Note high degree of similarity even in variable sections (cf. Fig. 4).

These comparisons make it very likely that the secondary and tertiary structures of the investigated A-Hc's are very similar. It appears that the micro-environment for each residue along the sequence is very constant and defended against evolutionary change.

Hopp and Woods (1981) have used such profiles to predict antigenic determinants. Antigenic determinants are on the surface of proteins and "are frequently found in regions that have an unusually high degree of exposure to the solvent" - such regions should correspond to the peaks in hydrophilicity profile. Figure 6 gives an example. Peaks marked by a bar would be listed as presumptive antigenic sites. However, if these positions are compared with the tertiary structure of Panulirus Hc (Gaykema et al., 1984) it becomes evident that some of the very hydrophilic sections are not on the surface of the molecule but in the interior. Of course, such a statement must be viewed with caution until the tertiary structure is known to more detail.

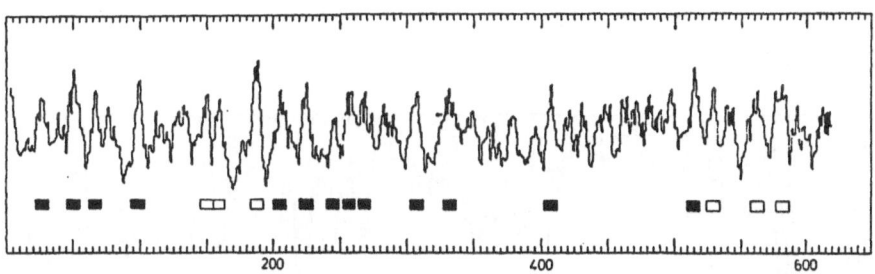

FIG. 6. Hydrophilicity profile (Hopp and Woods, 1981) of Eurypelma Hc subunit e with prediction of antigenic sites. Filled bars indicate hydrophilic sections on the surface of the molecule, while open bars indicate sections which are in the interior of the molecule and not accessible to antibodies.

The degree of homology, indicated above for the whole subunits, varies for the different parts of the chain. Table 1 shows a comparison of Eurypelma subunit a with 4 other cheliceratan subunits, grouped by the N-terminal (pos. 1-150), the central (pos. 151-370) and the C-terminal (pos. 371-623) parts. It can be seen that in each section similar values are obtained in the four comparaisons, but that the overall level of identity is different between the sections. There is one outstanding figure: 52 per cent identity for Eurypelma a vs. Limulus II. These two subunits occupy equivalent positions in the quaternary structure and they show a clear immunological crossreaction (Kempter et al., 1985). Another way to look at such similarities or dissimilarities is by constructing 'homology profiles' (Henschen et al., 1982) in a way similar to the hydrophilicity profiles. We have done this for two pairs of subunits: Eurypelma a/Limulus II (interspecific; equivalent subunits) and Eurypelma a/Eurypelma d (intraspecific; non-equivalent subunits). The comparison of both profiles (Fig. 7) shows that the pair Eury a/Lim II[*] shows as much or more homology than the pair Eury a/Eury d, in spite

[*] Eury: Eurypelma californicum; Lim.: Limulus polyphemus

66

TABLE 1. Per cent identical amino acids in the N-terminal (pos.[*] 1-150), central (pos. 151-370) and C-terminal (pos. 371-623) sections of cheliceratan hemocyanins. Eurypelma Hc subunit a is compared with Eurypelma d and e, and with two horseshoe crab Hc subunits. In the quaternary structure, Eury a is equivalent to Limulus II. Counting is according to Schartau et al. (1983). Note that the sequence of Eury a is incomplete, hence the lower number of amino acids compared in each section. Data from Metzer (1984).

Eury a versus	Pos. 1-150		Pos. 151-370		Pos. 371-623	
	Residues compared	Identity (%)	Residues compared	Identity (%)	Residues compared	Identity (%)
Eury d	127	41.7	219	65.3	224	54.5
Eury e	127	39.4	219	68.5	224	49.6
Lim II	127	52.0	219	70.3	214	48.6
Tachy α	127	42.5	219	66.7	224	48.2

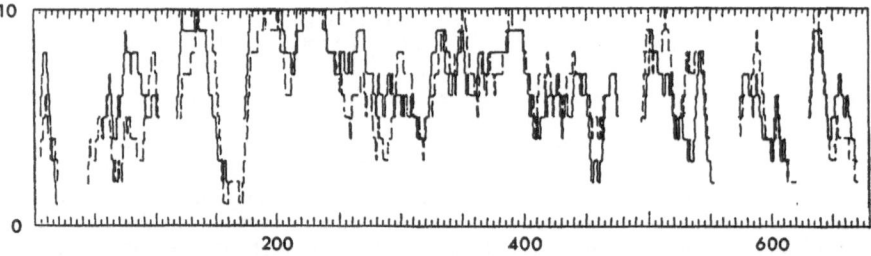

FIG. 7. Two "homology profiles" constructed for Hc subunits Eurypelma a / Limulus II (solid line) and Eurypelma d / Eurypelma a (dashed). Number of identical residues counted for groups of 10. Gaps indicate sections not yet sequenced. Note high degree of homology for the equivalent subunits Eury a and Lim II in spite of early divergence of Xiphosura from Arachnida.

of some 500 millions years of evolution since the divergence of the Arachnida from the Xiphosura. Those regions of significantly greater homology in the Eury a/Lim II comparison are supposedly engaged in intersubunit contacts. This is indeed very suggestive for pos. 65-87 and pos. 165-172 which are loops at the bottom and at the convex back surface of the subunit (Fig. 8).

It is most fortunate that only a few months after the completion of the first amino acid sequences also the first three-dimensional structure of an arthropodan hemocyanin was determined at 3.5 Å resolution (Gaykema et al., 1984). The complete α-carbon chain could be constructed, disulfide bridges located and sequenced sections identified. Most important, the binuclear copper site could be accurately described. The subunit consists of three

[*] pos.: position

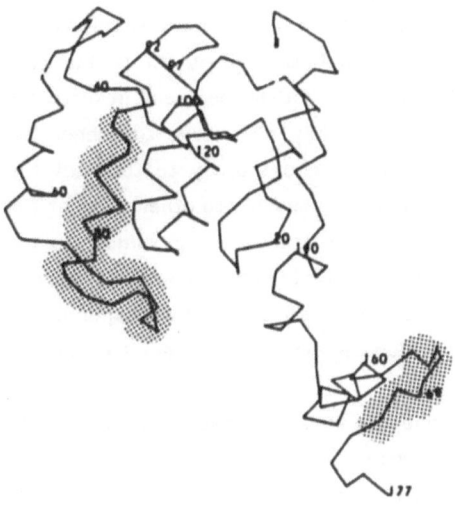

FIG. 8. α-Carbon chain of Domain 1 in Panulirus interruptus hemocyanin (Gaykema et al., 1984). The N-terminus is at the upper right hand corner. The shaded areas indicate sections where the sequences of Limulus II and Eurypelma a (equivalent in the quaternary structure) show greater homology than Eurypelma d and Eurypelma a. These regions might function in specific intersubunit contacts.

domains. Domain 1 (pos. 1-177) is mainly helical and has an appendix which reaches underneath Domain 2. Domain 2 (pos. 178-400) contains the copper-binding site about in its center; it is also mainly helical. Domain 3 is very different; it features a β-barrel consisting of 7 strands, and sends out two long loops extending past Domain 2 and making contact even with Domain 1.

We will procede now to discuss some remarkable features of the primary structure in relation to the tertiary structure. If we compare crustacean and cheliceratan Hc's there appear to be three larger deletions in the latter (or insertions in the former): 5 residues at the N-terminus, 21 residues between pos. 22 and 42, and about 8 residues (variable) in the third domain (pos. 564-573). All these are at the surface of the molecule. The large deletion at pos. 22-42 is an α-helix which bulges out from the side of Domain 1 (the 'front end' of the subunit) and which is not involved in intersubunit contacts. Its excision could not have had any major consequences except for the need to make the underlying structures more hydrophilic. Removal of pos. 564-573 must probably have altered the contacts between Domains 3 and 1, by shortening one of the two long loops of Domain 3.

The first domain is mainly variable; a conservative region begins around pos. 110 at an α-helix (1.6) which is entirely buried in the interior and hence strongly hydrophobic. It may be important in making contact to one of the helices which provide for the active site. Close to the end of Domain 1 there is the carbohydrate attachment site in Panulirus Hc (Vereijken et al., 1982).

Domain 2 contains the binuclear copper site. The two copper atoms are ligated by three His residues each, the six His being contributed by four α-helices. There is no 'bridging ligand' as had been proposed on the basis of spectroscopic evidence (Himmelwright et al., 1980). The very remarkable sequence -His-His-Trp-His-Trp--His- which had been predicted as part of the active site (Schneider et al., 1982) was shown to comprise the first two ligands of copper A. The third ligand is a histidine 27 residues downstream. A similar constellation is found around pos. 351-355 and pos. 393 for copper B. In both regions the amino acid sequence is extremely conserved. It is remarkable that most (50-80 per cent in different subunits) of the Trp residues are found in these sections. They may serve to anchor the active site helices in the surrounding hydrophobic regions and thereby to transmit the movements of the copper atoms during oxygenation and deoxygenation to other parts of the subunit. Domain 2 is divided by a steep cleft running down from the top almost to the bottom surface and lodging the two copper atoms. This cleft would probably widen or narrow along with the copper movement. The proximity of the Trp residues to the copper atoms explains the strong effect of oxygenation in Trp fluorescence by non-radiative energy transfer (Loewe, 1978).

Two other groups of amino acids near the active site strike our eyes. One is a cluster of basic residues (pos. 217-223), especially pronounced in the cheliceratan Hc's and embraced by two completely conserved glycines. These are at the top of Domain 2 in the loop between helices 2.1 and 2.2. No proposal with respect to their function can presently be made. Another interesting cluster is around pos. 380 - a group of hydroxy amino acids especially well conserved in the Chelicerata. This is at the bottom of the deep cleft in Domain 2; it may bind water in the interior of the protein.

Space does not permit to discuss the rest of the polypeptide chain although there are many interesting features. The last one worth to be pointed out is the frequency by which His residues occur close to the C-terminus. Again, no explanation based on experimental evidence can be offered at present but these residues may serve in one of the mechanisms by which hemocyanin function is modulated by pH.

Within about two and a half years we have witnessed a complete breakthrough in the structure of arthropodan hemocyanins: the quaternary, the primary and the tertiary plus secondary structures have basically all been clarified. The progress has been so rapid that we are not yet able to fully evaluate all data. The comparison of the results at the four levels of molecular structure will give us new insights with respect to the architecture and the physical function of the whole, hexameric or multi-hexameric molecule. At the same time, new problems can be addressed. One of them is the relation between arthropodan and molluscan hemocyanins. There are such profound differences between the two that many workers have hesitated to believe in a common ancestry. At these meeting the first structural data are presented (Schneider et al., cf. the Abstracts volume) which show a common origin of these two respiratory proteins but also the extremely great distance. Another challenging problem is the way by which cooperative interaction between the subunits is achieved, the mechanism of the Bohr effect and of modulation by other effectors both inorganic and organic. In figurative expression: We have just landed on a new continent, but we have not yet crossed it.

ACKNOWLEDGMENTS

We are most grateful for the help provided by Dr. R. Paul (computer programming and evaluation), by Ms H. Storz and Ms G. Feldmaier (technical assistance) and Ms I. Krella (type-script). Financial support to the author's laboratory came from the Deutsche Forschungsgemeinschaft (Li 107, Scha 317, Schn 226).

REFERENCES

Braunitzer G, Gehring-Müller R, Hilschmann N, Hilse K, Hobom G, Rudloff V, Wittmann-Liebold B (1981) Die Konstitution des normalen adulten Humanhämoglobins. Hoppe-Seyler's Z. Physiol. Chem. 325: 283-286

Busselen P (1970) The electrophoretic heterogeneity of Carcinus maenas hemocyanin. Arch. Biochem. Biophys. 137: 415-420

Chang JY, Brauer D, Wittmann-Liebold B (1978) Miroc sequence analysis of peptides and proteins using 4-NN-dimethylaminoazobenzene 4'-isothiocyanate/phenylisothiocyanate double coupling method. FEBS Lett. 93: 205-214

Ellerton HD, Ellerton NF, Robinson HA (1983) Hemocyanin - a current perspective. Progr. Biophys. Molec. Biol. 41: 143-248

Gaykema WPJ, Hol WGJ, Vereijken JM, Soeter NM, Bak HJ, Beintema JJ (1984) 3.2 Å Structure of the copper-containing, oxygen-carrying protein, Panulirus interruptus hemocyanin. Nature (London) 309: 23-29

Henschen A, Lottspeich F, Kehl M, Southan C, Lucas J (1982) Structure-function-evolution relationship in fibrinogen. In: Henschen A, Graeff H, Lottspeich F (eds) Fibrinogen -Recent Biochemical and Medical Aspects. Walter de Gruyter, Berlin, New York, pp. 67-82

Himmelwright RS, Eickman NC, Lubien CD, Solomon EI (1980) Chemical and spectroscopic comparison of the binuclear copper active site of mollusc and arthropod hemocyanins. J. Amer. Chem. Soc. 102: 5378-5388

Hopp TP, Woods KR (1981) Prediction of protein antigenic determinants from amino acid sequences. Proc. Natl. Acad. Sci. USA 78: 3824-3828

Jollès J, Jollès P, Lamy J, Lamy J (1979) Structural characterization of seven different subunits in Androctonus australis hemocyanin. FEBS Lett. 106: 289-291

Kempter B, Markl J, Gebauer W, Brenowitz M, Bonaventura C, Bonaventura J (1985) Immunological correspondence between arthropod hemocyanin subunits. II. Xiphosuran (Limulus) and spider (Eurypelma, Cupiennius) hemocyanin. Hoppe-Seyler's Z. Physiol. Chem. (submitted)

Klarman A, Daniel E (1981) Structural basis of subunit heterogeneity in arthropod hemocyanins. Comp. Biochem. Physiol. 70B: 115-123

Klippenstein GL, Holleman JW, Klotz IM (1968) The primary structure of Golfingia gouldii hemerythrin. Order of peptides in fragments produced by tryptic digestion of succinylated hemerythrin. Complete amino acid sequence. Biochemistry 7: 3868-3878

Lamy J, Richard M, Goyffon M (1970) Sur les modifications des électrophorégrammes en gel de polyacrylamide des protéines de l'hémolymphe des Scorpions Androctonus australis (L.) et Androctonus mauretanicus (Pocock), provoqués par la congélation. C.R. Acad. Sci. Paris, série D, 270: 1627-1630

Lamy J, Lamy J, Weill J, Bonaventura J, Bonaventura C, Brenowitz M (1979) Immunological correlates between the multiple hemocyanin subunits of Limulus polyphemus and Tachypleus tridentatus. Arch. Biochem. Biophys. 196: 324-339

Lamy J, Bijlholt MMC, Sizaret P-Y, Lamy J, van Bruggen EFJ (1981) Quaternary structure of scorpion (Androctonus australis) hemocyanin. Localization of subunits with immunological methods and electron microscopy. Biochemistry 20: 1849-1856

Lamy J, Sizaret P-Y, Lamy J, Feldmann R, Bonaventura J, Bonaventura C (1982) Preliminary report on the quaternary structure of Limulus polyphemus hemocyanin. Life Chemistry Reports, suppl. 1: 47-50

Lamy J, Compin S, Lamy J (1983) Immunological correlates between multiple isolated subunits of Androctonus australis and Limulus polyphemus hemocyanins: an evolutionary approach. Arch. Biochem. Biophys. 223: 584-603

Linzen B (1983) Subunit heterogeneity in arthropodan hemocyanins. Life Chemistry Reports suppl. 1: 26-38

Loewe R (1978) Hemocyanins in spiders. V. Fluorimetric recording of oxygen binding curves, and its application to the analysis of allosteric interactions in Eurypelma californicum hemocyanin. J. Comp. Physiol. 128: 161-168

Markl J, Strych W, Schartau W, Schneider H-J, Schöberl P, Linzen B (1979) Hemocyanins in spiders. VI. Comparison of the polypeptide chains of Eurypelma californicum hemocyanin. Hoppe-Seyler's Z. Physiol. Chem. 360: 639-650

Markl J, Kempter B, Linzen B, Bijlholt MMC, van Bruggen EFJ (1981) Hemocyanins in spiders. XVI. Subunit topography and a model of the quaternary structure of Eurypelma hemocyanin. Hoppe-Seyler's Z. Physiol. Chem. 362: 1631-1641

Markl J, Decker H, Linzen B, Schutter WG, van Bruggen EFJ (1982) Hemocyanins in spiders. XV. The role of the individual subunits in the assembly of Eurypelma hemocyanin. Hoppe-Seyler's Z. Physiol. Chem. 363: 73-87

Markl J, Gebauer W, Runzler R, Avissar I (1984) Immunological correspondence between arthropod hemocyanin subunits. I. Scorpion (Leiurus, Androctonus) and spider (Eurypelma, Cupiennius) hemocyanin. Hoppe-Seyler's Z. Physiol. Chem. 365: 619-631

Metzger W (1984) Zur Primärstruktur der Untereinheit a des Hämocyanins aus der Vogelspinne Eurypelma californicum. Diploma Thesis, University of Munich

Nemoto T, Takagi T (1983) Sequence of Tachypleus tridentatus hemocyanin. Reported at the 56th Ann Meeting Jap Biochem Soc, Sept. 29 - Oct. 2.

Nozaki Y, Tanford C (1971) The solubility of amino acids and two glycine peptides in aqueous ethanol and dioxane solutions. Establishment of a hydrophobicity scale. J. Biol. Chem. 246: 2211-2217

Schartau W, Eyerle F, Reisinger P, Geisert H, Storz H, Linzen B (1983) Hemocyanins in spiders. XIX. Complete amino acid sequence of subunit d from Eurypelma californicum hemocyanin and comparison to chain e. Hoppe-Seyler's Z. Physiol. Chem. 364: 1383-1409

Schneider H-J, Markl J, Schartau W, Linzen B (1977) Hemocyanins in spiders. IV. Subunit heterogeneity of Eurypelma (Dugesiella) hemocyanin, and separation of polypeptide chains. Hoppe-Seyler's Z. Physiol. Chem. 358: 1133-1141

Schneider H-J, Illig U, Müller E, Linzen B, Lottspeich F, Henschen A (1982) Hemocyanins in spiders. XVII. A presumptive active-site sequence of arthropodan hemocyanins. Hoppe-Seyler's Z. Physiol. Chem. 363: 487-492

Schneider H-J, Drexel R, Feldmaier G, Linzen B, Lottspeich F, Henschen A (1983) Hemocyanin in spiders. XVIII. Complete amino-acid sequence of subunit e from Eurypelma californicum hemocyanins. Hoppe-Seyler's Z. Physiol. Chem. 364: 1357-1381

Sizaret P-Y, Frank J, Lamy J, Weill J, Lamy J (1982) A refined quaternary structure of Androctonus australis hemocyanin. Eur. J. Biochem. 127: 501-506

Sullivan B, Bonaventura J, Bonaventura C, Godette G (1976) Hemocyanin of the horseshoe crab, Limulus polyphemus. Structural differentiation of the isolated components. J. Biol. Chem. 251: 7644-7648

Takagi T, Nemoto T (1980) Tachypleus tridentatus hemocyanin. Separation and characterization of monomer subunits and studies of sulfhydryl groups. J. Biochem. (Tokyo) 87: 1785-1793

van Holde KE, Miller KI (1982) Hemocyanins. Quart. Rev. Biophys. 15: 1-129

Vereijken JM, Schwander EH, Soeter NM, Beintema JJ (1982) Limited proteolysis of the 94000-dalton subunit of Panulirus interruptus hemocyanin; the carbohydrate attachment site. Eur. J. Biochem. 123: 283-289

Yokota E, Riggs AF (1984) The structure of the hemocyanin from the horseshoe crab, Limulus polyphemus. The amino acid sequence of the largest cyanogen bromide fragment. J. Biol. Chem. 259: 4739-4749

Zwilling R, Dörsam H, Torff H-J, Rödl J (1981) Low molecular mass protease: evidence for a new family of proteolytic enzymes. FEBS Lett. 127: 75-78

Quaternary Structure of Arthropod Hemocyanins

J. LAMY, J. LAMY, P.-Y. SIZARET, P. BILLIALD, G. MOTTA

I. INTRODUCTION

Hemocyanins, the blue respiratory pigments of many Arthropods, are generally considered as resulting from successive dimerizations of a building block composed of six kidney-shaped (70-75 kDa) subunits. This basic structural unit, called hexamer or (1x6)-mer, is a constituent of both crustacean and cheliceratan hemocyanins. Crustaceas contain (1x6)-meric hemocyanins in the spiny lobster Panulirus interruptus or in the shrimp Penaeus setiferus, (2x6)-mers in most species, and tetrahydral (4x6)-mers in some Thalassinid shrimps belonging to the genus Callianassa and Upogebia. Chelicerates which are considered more primitive than Crustaceas, at least from a paleontological point of view, possess more complex hemocyanins. Actually, hemocyanins of all the still living horseshoe crabs are (8x6)-mers, those of scorpions, primitive spiders, Uropygia and Amblypygia are flat (4x6)-meric structures while those of modern spiders are mostly (2x6)-mers. A more complete description of the phylogenic distribution of hemocyanins is found in recent reviews (Ellerton et al., 1983; van Holde and Miller, 1982).

This paper deals with the determination of the quaternary structures of the (8x6)-meric hemocyanin of the horseshoe crab Limulus polyphemus, and of the (4x6)-meric hemocyanins of the scorpion Androctonus australis and the spider Eurypelma californicum. The determination of a quaternary structure is not a single step work. Indeed, before assigning to each copy of each polypeptide chain a location within the whole molecule, it is necessary to know, first, the architectural organization of the subunits and, second, the number of copies of the various polypeptide chains in the oligomer.

II. DETERMINING THE NUMBER OF COPIES OF EACH SUBUNIT IN OLIGOMER

The most complex cheliceratan hemocyanins are composed of at least 7 to 8 different polypeptide chains with molecular weight about 70-75 kDa. Recently, Linzen (1983) has reviewed this question in a paper to which the reader is referred for more detailed information.

The subunit heterogeneity of the 3 chelicerate hemocyanins which are the matter of this paper was established on chromatographic, electrophoretic, immunologic, and amino acid sequences bases.

After removal of the calcium ions by EDTA or by extensive dialysis, the whole molecules are dissociated at neutral pH in the presence of 1 M urea, or at alcaline pH. Then, dissociated hemocyanins are fractionated by various techniques such as ion-exchange chromatography, polyacrylamide gel electrophoresis, isoelectrofocusing or crossed immuno-electrophoresis. Figure 1 shows an example of crossed immunoelectrophoresis of dissociated hemocyanin from A. australis against an homologous antiserum. The nine peaks correspond to eight immunologically distinct monomeric subunits (termed Aa 2, 3A, 3B, 3C, 4, 5A, 5B, and 6) and to a dimeric subunit (Aa 3C-5B) composed of one copy of subunit Aa 3C and one copy of subunit Aa 5B (Lamy et al., 1979a). All the monomeric subunits were shown later to have different N-terminal sequences of amino acids (Jollès et al., 1979). However, in the case of L. polyphemus hemocyanin, strong discrepancies occurred between electrophoretic and immunological heterogeneity. Thus Brenowitz et al. (1981) found not less than 16 electrophoretic species, where Lamy et al. (1979b) had observed only 8 immunologically different subunits. This difference implies that some of the electrophoretic subunits are immunologically identical. For example, the subunits, termed Lp I, I', and I" by Brenowitz et al. which had identical N-terminal sequences, intramolecular locations, structural and functional properties (Brenowitz, 1984a; Lamy et al., 1983a), could be distinguished only by their electrophoretic mobilities in polyacrylamide gel electrophoresis. Therefore, it seems reasonable to now consider that immunologically pure subunits are "true" subunits because their N-terminal sequences are different, but that they may exhibit a microheterogeneity. This minor problem will probably not be solved until the complete amino acid sequence of immunologically pure subunits demonstrates whether or not they are chemically pure.

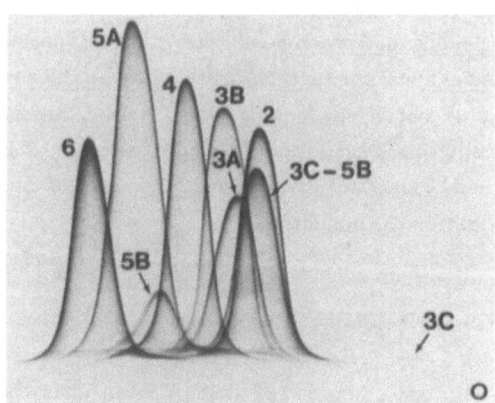

FIG. 1. Crossed immunoelectrophoresis of dissociated Androctonus australis hemocyanin against an homologous antiserum. The peaks are labelled according to the subunit nomenclature of Lamy et al. (1979b).

The determination of the number of copies of the various subunits in the whole molecules also raises some methodological problems. The method consists of a dissociation of the whole molecule followed by an assay of each subunit in the mixture of free subunits.

If the dissociation step is usually performed easily, the assay step is more difficult. Two methods are used in practice: first, a densitometric integration of electrophoretic or chromatographic elution profiles and, second, an immunological determination. The densitometric method is more or less imprecise because the overlaps between the elution fractions lead to an overestimation of the less abundant subunits. Conversely, the immunoassay does not require a preliminary separation of the components of the mixture and gives good results but it requires that one antigen is determined at a time and that the antibodies do not bind to any other antigen. The problem is that cross-reactivities generally occur between the subunits of one hemocyanin (Lamy et al., 1983b). Therefore, it is incorrect to make a planimetric integration of overlaping subunit peaks produced in crossed immunoelectrophoresis against an anti-dissociated hemocyanin serum as preconized by Markl et al. (1981a), and Brenowitz et al. (1984a). This procedure can give important over- or underestimations depending on cross-reactivities between subunits present in overlaping peaks. Three methods may be correctly used: radioimmunoassay, enzyme immunoassay, and rocket immunoelectrophoresis, provided that the antibody preparation is rigorously specific of the subunit to be determined. Such a high specificity may result from a careful selection of monoclonal antibodies or from a series of immunoadsorptions of polyclonal antibodies to the cross-reacting subunits present in dissociated hemocyanin.

TABLE 1. Approximate number of copies per (4x6)-mer (rounded to the nearest integer) of the various subunits of Androctonus australis (Aa), Eurypelma californicum (Ec), and Limulus polyphemus (Lp) hemocyanin

Aa Lamy et al. 1981a		Ec Markl et al. 1981b		Lp Lamy et al. 1983a	
2	4	a	4	I	3
3A	2	b	2	II	4
3B	2	c	2	IIA	1
3C	2	d	4	IIIA	4
4	4	e	4	IIIB	4
5A	4	f	4	IV	4
5B	2	g	4	V	2
6	4			VI	2

Table 1 shows the subunit composition of 3 chelicerate hemocyanins. To make the results more understandable, the numbers of copies have been rounded to the nearest integer. The subunit composition of L. polyphemus hemocyanin asks the important question of the homogeneity of the native molecule population. Indeed, the data of Table 1 demonstrate that the (4x6)-meric half-molecule cannot be composed of two identical dodecamers (1 copy of subunit Lp IIA per (4x6)-mer)); this means either that two (or more) types of dodecameric quarters of molecules exist or that subunit Lp IIA and perhaps other subunits are randomly distributed in the (4x6)-mers.

III. RESOLVING THE ARCHITECTURE OF THE WHOLE MOLECULES

How are 24 and 48 subunits assembled in A. australis or E. californicum and Limulus polyphemus hemocyanins respectively ? This question is less complex than it could appear at first. Indeed, the (8x6)-meric hemocyanin of L. polyphemus dissociates under mild conditions into (4x6)-mers impossible to distinguish in the electron microscope from those of A. australis and E. californicum, even by the most sophisticated technique of image analysis (Bijlholt et al., 1982). On the other hand, the three-dimensional structure of all the Arthropod hemocyanin subunits seems roughly similar. Recently, Gaykema et al. (1984) determined, by X-ray crystallography at a 3.2 Å resolution, the three-dimensional structure of Panulirus interruptus subunits and, evidence exists that subunit Lp II from L. polyphemus and subunits Aa 2 and Aa 4 from A. australis may have similar structures (Fearon et al., 1983; Magnus, 1983). Therefore, the problem is reduced to understanding how 24 kidney-shaped subunits of the P. interruptus type are arranged in a (4x6)-mer. In a first approach, advantage was taken of the E.M. views of the various Arthropod hemocyanins. As shown in figures 2 and 3, all the arthropod hemocyanins are supposed to be built of hexameric building blocks similar to P. interruptus hemocyanin. In this structure, called a trigonal antiprism, the 6 subunits are disposed in two layers of 3, the upper layer being rotated 60° around the 3-fold axis. Two copies of this hexamer are further assembled into a dodecamer similar to crustacean hemocyanins, then two such dodecamers are put side by side to make a (4x6)-mer (Lamy et al., 1981b; Lamy et al., 1982). Finally, two (4x6)-mers are superimposed, then the upper one is rotated approximately 45° producing a (8x6)-mer in fairly good agreement with the five E.M. views of L. polyphemus hemocyanin (Lamy et al., 1982).

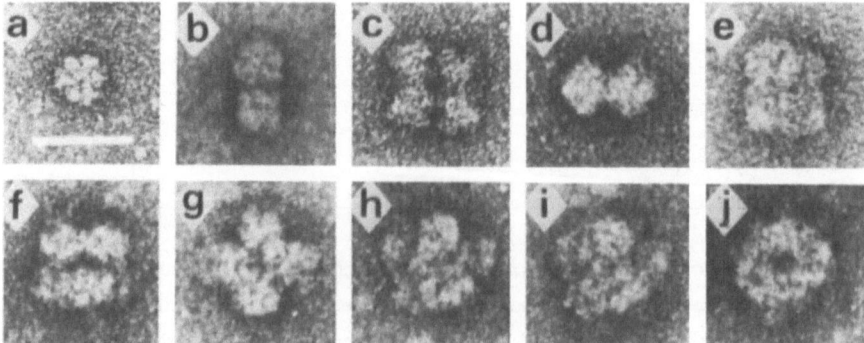

FIG. 2. The various E.M. views of Arthropod hemocyanins: a) Hexamer from Panulirus interruptus; b) dodecamer from Homarus americanus; c) top view; d) side view; e) 45°-view from (4x6)-mers from Androctonus australis; f) bowtie view; g) cross view; h) symmetric pentagon; i) asymmetric pentagon; j) ring view from the (8x6)-mers of Limulus polyphemus. The bar is 25 nm.

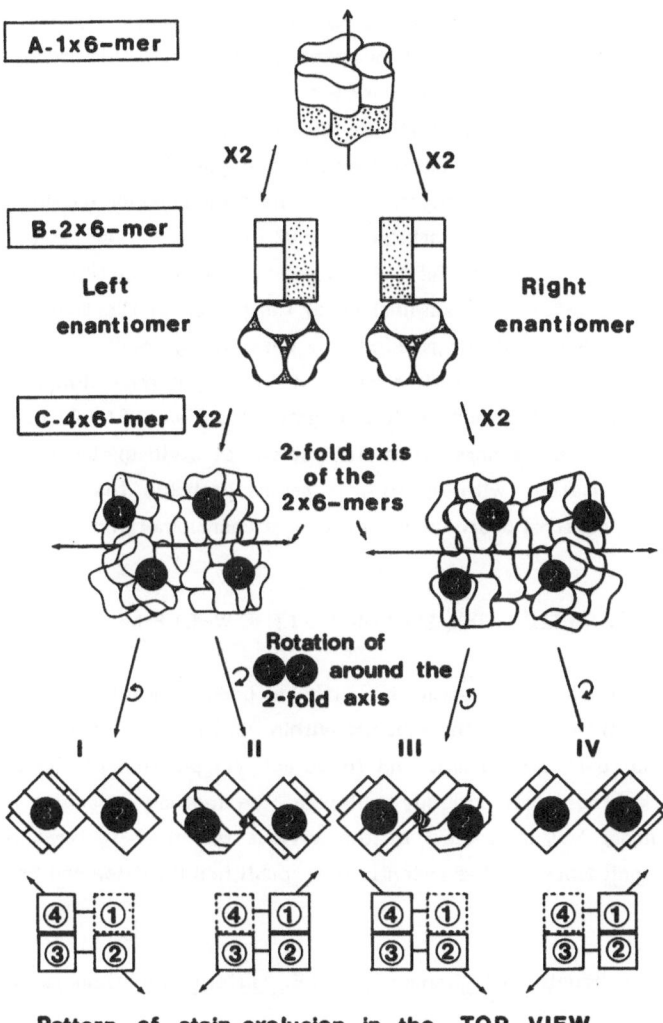

FIG. 3. Schematic architectural organization of Arthropod hemocyanins: A) Hexamers of the <u>Panulirus interruptus</u> type. B) Dodecamers of the left and right enantiomorphic types. C) The four (4x6)-mers resulting from the introduction of the rocking effects into dodecamers of the left and right types. □ Strong stain exclusion; ⋮ reduced stain exclusion. ◄──────► Rocking axis.

To take into account the important discovery of the rocking effect (van Heel and Frank, 1981), the concept of the (2x6)-mer enantiomorph was proposed by Lamy <u>et al.</u> (1982). The rocking effect describes the fact that the (4x6)-mer molecule in its top view does not stand on the support on four hexamers but on only three. The result is that the molecule is shaky and "rocks" around an axis joining the lower right and the upper left hexamers. The enantiomorph concept, summarized in figure 3, states that there are two and only two ways of assembling two identical hexamers, to produce a dodecamer similar to those of Crustacea, and that the two dodecamers resulting from this process are a pair of enantiomorphs. By convention, the dodecamers resulting from a 90° rotation of the upper hexamer to the right and to the left are designated RIGHT and LEFT enantiomers, respectively. The orientation of the rocking axis to the upper left was first interpreted as a consequence of the enantiomorphic asymmetry of the dodecamer composing the (4x6)-mers (left type). However, at the EMBO workshop of 1982 in Leeds, M. van Heel presented a new E.M. view of the (4x6)-mer, called "45°-view", which was incompatible with the proposed model (Sizaret et al., 1982). A reexamination of the choice of the enantiomer led to the conclusion that the direction of the rocking axis does not by itself allow a discrimination between the enantiomorphs as was first believed. In fact, it only allows the rejection of models II and IV in figure 3, but models I (left enantiomorph) and III (right enantiomorph) are still both acceptable. Furthermore, with the pattern of stain exclusion of the 45°-view appearing hardly compatible with model I, it seems probable that all the Chelicerate hemocyanins are composed of dodecamers of the right enantiomorphic type.

IV. LOCALIZING THE SUBUNITS IN THE WHOLE MOLECULES

As the most complex hemocyanins have not yet been crystallized, X-ray crystallography could not be used to localize the subunits within the whole molecule. Another approach was to bind a tag onto the subunit and to detect the position of the tag by electron microscopy. If the tag is an antibody molecule, the method is called immunoelectron microscopy (immunoE.M.). A review of the various technical features of immunoE.M. applied to the E. coli ribosome has recently been published (Stöffler and Stöffler-Meilicke, 1984).

A. Quaternary Structure of A. australis and E. californicum Hemocyanins

The first probes were carried out with polyclonal IgG's specific for the various subunits of A. australis hemocyanin. The results were disappointing because of the immunoprecipitation. Indeed, an incubation of subunit-specific IgG's with hemocyanin molecules, possessing several copies of each subunit, produced a three-dimensional network leading to a precipitation. The problem was how to prevent the precipitation without losing the high specificity of the antibody binding to the antigen. A splitting of the divalent IgG molecule into monovalent Fab fragments and a purification of the immuno complexes by gel filtration allowed the observation of many hemocyanin molecules labelled by Fab fragments. The position of the Fab fragments pointing out the contour line of the hemocyanin molecule indicated the topographical location of the subunit. Figure 4 shows some soluble immuno complexes obtained with A. australis hemocyanin and a preparation of Fab fragments specific for subunit Aa 3B. The position of the Fab fragments demonstrates that subunit

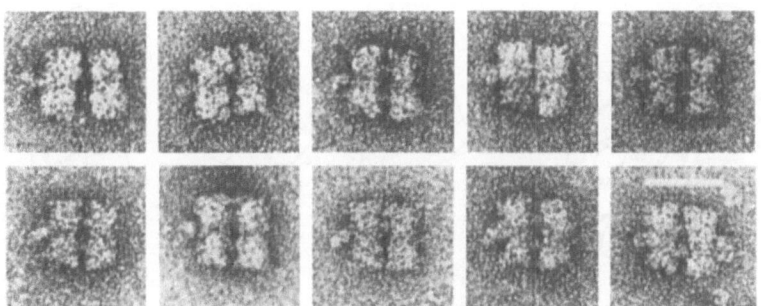

FIG. 4. Immunolabelling of subunit Aa 3B in Androctonus australis hemocyanin. Selected views of soluble immuno complexes obtained with anti-Aa 3B Fab fragments. The bar is 25 nm.

Aa 3B is located near the middle of the lateral edge in the top view. All the eight polypeptide chains of A. australis were roughly localized by this technique (Lamy et al., 1981b). However, some doubt remained with respect to the relative positions of subunits Aa 3A, 3B, 3C, and 5B. Indeed, subunits Aa 3A and 3B produced the same type of immuno complexes, as well as subunits Aa 3C and 5B. As indicated in Table 1, these four subunits are present in two copies per (4x6)-mer (1 copy per half-dodecameric molecule). Furthermore, within each dodecamer, Aa 3A and Aa 3B were located in different hexamers as well as Aa 3C and Aa 5B. Therefore, the problem was to determine which subunit of each pair (Aa 3A,3B and Aa 3C,5B) are located in the same hexamer.

The ambiguity was solved by double immunolabelling experiments. The principle of the method was to label the hemocyanin molecule by various binary mixtures of Fab fragments specific for the four above-mentioned subunits. Figure 5 shows an example of the results, which demonstrates that, within each dodecameric half-molecule, subunits Aa 3A and Aa 3C are located in the same hexamer while subunits Aa 3B and Aa 5B are in the other hexamer (Sizaret et al., 1982). This result and other results of immunolabelling experiments were included in a model of (4x6)-mer, shown in figure 6, composed of 2 copies of the right dodecameric enantiomorph.

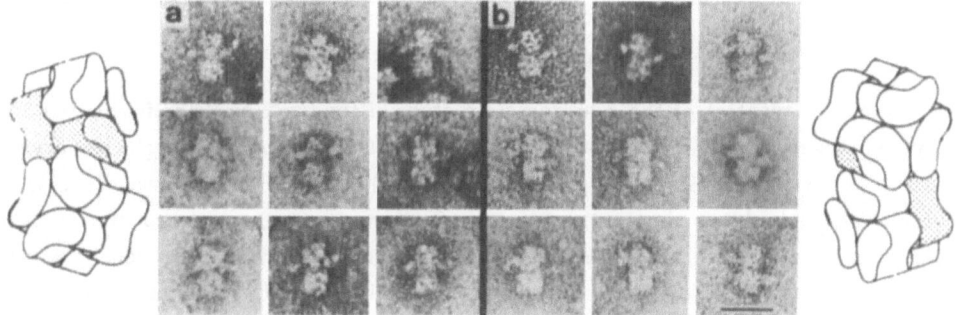

FIG. 5. Double immunolabelling of A. australis hemocyanin by a mixture of Fab fragments specific for subunits : a) Aa 3B + Aa 3C and b) Aa 3B + Aa 5B. The bar is 25 nm.

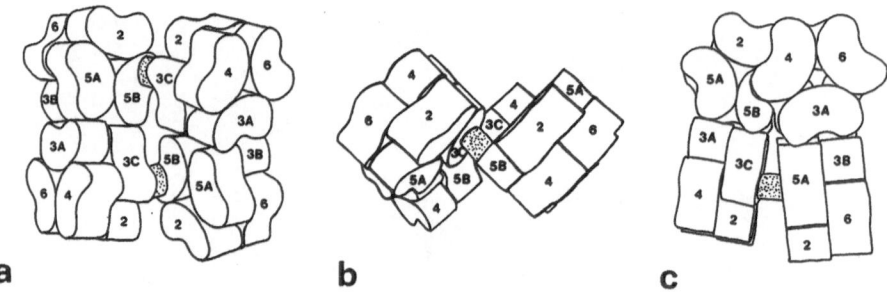

FIG. 6. Model of quaternary structure of Androctonus australis hemocyanin. The (2x6)-mer enantiomorph used to build the model is of the right type: a) Top view; b) Side view; c) 45°-view.

The same method was further applied to the intramolecular localization of the spider E. californicum. As shown in Table 2, similar results were obtained with the exception that the two positions occupied by subunits Aa 3A and Aa 3B in A. australis hemocyanin are filled in E. californicum by two copies of subunit Ec a (Markl et al., 1981b).

TABLE 2 - Intramolecular location of the subunits in the (4x6)-mers of Androctonus australis (Aa), Eurypelma californicum (Ec), and Limulus polyphemus (Lp) determined by immuno-electronmicroscopy

Location	E.M. view	Aa	Ec	Lp
Corner	top	6	e	I,IIA
Middle of the lateral edge	top	3A,3B	a	II
Middle of the end of the (2x6)-mer	top	4	g	IIIA
End of the (2x6)-mer near the cleft	top	2	b	IIIB
Top/bottom edge	side	5A	d	IV
Interdodecamer bridge	top	3C,5B	bc	V,VI

B. Quaternary Structure of L. polyphemus Hemocyanin

The determination of the quaternary structure of the (4x6)-meric half-molecule of L. polyphemus hemocyanin was not a simple repetition of the work on A. australis hemocyanin. Indeed, as shown above, progressively increasing the pH in the absence of calcium, successively dissociates the (8x6)-mers into (4x6)-mers, (2x6)-mers and free subunits (Brenowitz et al., 1984b). However, the (4x6)-mers obtained under these conditions were very unstable and dissociated into subunits upon the binding of Fab fragments. To prevent this dissociation, the (4x6)-mers were first strengthened by dimethylsuberimidate, a bifunctional agent which crosslinks the subunits without strongly altering their immunolo-gical properties. The crosslinked (4x6)-mers remained stable under dissociating conditions and produced soluble immuno complexes resembling those previously obtained with A. australis hemocyanin (Billiald et al., 1983). The intramolecular location of the subunits in L. polyphemus half-molecules is indicated in Table 2. Their topographical position in the model can be deduced from those of A. australis subunits in figure 6.

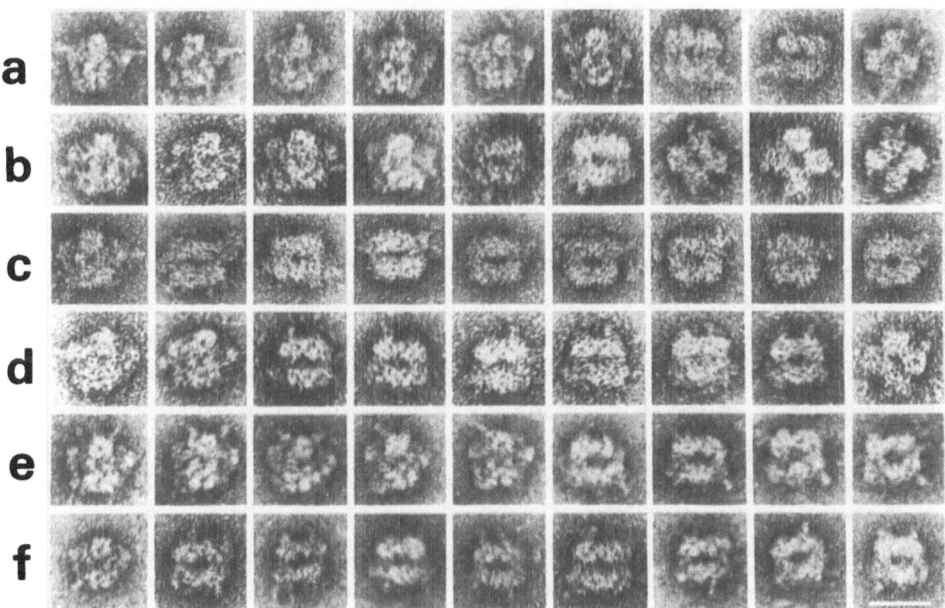

FIG. 7. Gallery of (8x6)-meric Limulus polyphemus crosslinked hemocyanin labelled with antisubunit Fab fragments. a-f) Labelling with anti Lp I, II, IIA, IIIA, IIIB, IV. The bar is 25 nm.

Crosslinked (8x6)-mers were also submitted to Fab labelling. However, the interpretation of the results was much more difficult because of the existence of the five E.M. views shown in figure 2. Selected images of the (8x6)-mers labelled by the various subunit-specific Fab fragments are shown in figure 7. They are in good agreement with the model of figure 8 built from four copies of the (2x6)-mer right enantiomer.

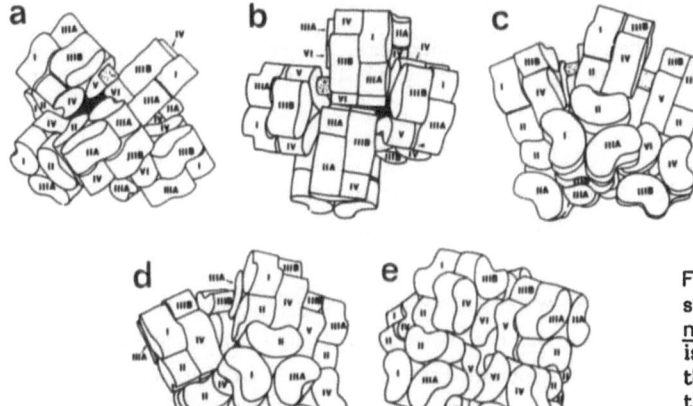

FIG. 8. Model of quaternary structure of Limulus polyphemus hemocyanin. The model is composed of four copies of the right enantiomorphic type: a) Bowtie, b) cross; c) pentagon; d) asymmetric pentagon; e) ring.

V. STUDYING THE STRUCTURAL ROLES OF SUBUNITS BY
REASSEMBLY OF OLIGOMERS AND IMMUNOLABELLING

The location of certain subunits in specific areas of the molecule suggests that they play specific structural and/or functional roles. For example, the position of subunit Aa 3C and 5B, Ec b and c, Lp V and VI in the area of the bridges joining the (2x6)-mers suggests that these subunits are involved in interdodecamer contacts. Furthermore, these subunits exhibit strong interspecies immunological cross-reactivities (Lamy et al., 1983b). Their implication is demonstrated by the following reassembly experiments.

First, a complete subunit mixture from A. australis hemocyanin, allowed to reassociate at neutral pH in the presence of calcium, produces (4x6)-mers resembling native hemocyanin. Secondly, if subunits Aa 3C and 5B are omitted from the reassembly mixture, the higher oligomer observed is a hexamer. Thirdly, the addition to the previous mixture of subunit Lp V-VI, having the same intramolecular location as subunit Aa 3C-5B restores the ability of the mixture to produce reassembled (4x6)-mers, suggesting that subunit Lp V-VI is capable of replacing subunit Aa 3C-5B in the A. australis system. In another experiment of this type, hybrid (4x6)-mers were reconstructed from the complete mixture of A. australis subunits depleted in subunits Aa 3A and 3B and supplemented in subunit Lp II. As shown in figure 9, an immunolabelling of these hybrid oligomers with anti-Lp II Fab fragments actually demonstrated that subunit Lp II replaces subunits Aa 3A and 3B. This was the first direct demonstration of the replacement stricto sensu of one subunit by another in a reconstructed oligomer (Lamy et al., 1983a). More detailed information about reassembly of oligomers from free subunits has been extensively published elsewhere (Bijlholt et al., 1979; Brenowitz et al., 1984a; Lamy et al., 1977; Lamy et al., 1980; Lamy et al., 1983a; Markl et al., 1982; van Bruggen et al., 1980).

FIG. 9. Selected views of reconstructed hybrid oligomers. The reaction mixture contained subunits Aa 2, 3C-5B, 4, 5A, 6, and Lp II. The same mixture without Lp II only produces hexamers. The bar is 25 nm.

VI. STRUCTURE - FUNCTION - EVOLUTION RELATIONSHIPS :
A STRUCTURAL APPROACH

The functional properties of the whole hemocyanin of A. australis, E. californicum, and L. polyphemus and of their isolated subunits have been studied. Although the 3 animals live in very different environments, it appears that the dimeric subunits involved in the interdodecamer contacts (Aa 3C-5B, Ec b-c, Lp V-VI) have a higher oxygen affinity than other subunits (Brenowitz et al., 1984 a,b; Lamy et al., 1980; Markl et al., 1981c). Furthermore, at least for A. australis and L. polyphemus, they have been better preserved by evolution than monomeric subunits. For example, it has been shown by factorial correspondence analysis of immunologic cross-reactivities that subunit Aa 3C-5B resembles subunit Lp V-VI more than it resembles other A. australis subunits and vice versa (Lamy et al., 1983b), suggesting that higher oxygen affinity is an important functional character.

Similarly, subunits Aa 5A and Lp IV, which occupy the same intramolecular position, have been preserved by evolution suggesting some important function (Lamy et al., 1983b). In fact, Lp IV exhibits a very unusual property among other hemocyanin subunits; its homohexamer is capable, in the presence of calcium, of cooperatively binding oxygen (Brenowitz et al., 1983). It has not yet been established whether or not subunit Aa 5A has the same property, and whether or not the cooperative oxygen binding of subunit Lp IV is the reason of its evolutionary preservation. However, the comparison of the functional and immunologic properties to the intramolecular location of the subunits leads one to think that the time when the relationship between structure and function will be clarified at a molecular level is not so far away. In such a perspective, the immunolabelling method has been improved using monoclonal instead of polyclonal antibodies. The preliminary results shown in figure 10 demonstrate that it is now possible to localize the label in a small area of a single subunit. With the first complete amino acid sequences (Schartau, 1983; Schneider et al., 1983) and the first three-dimensional structure (Gaykema et al., 1984) becoming now available, one can predict that this tool will help in the near future to detect with precision the conformational changes linked to the cooperative oxygen binding.

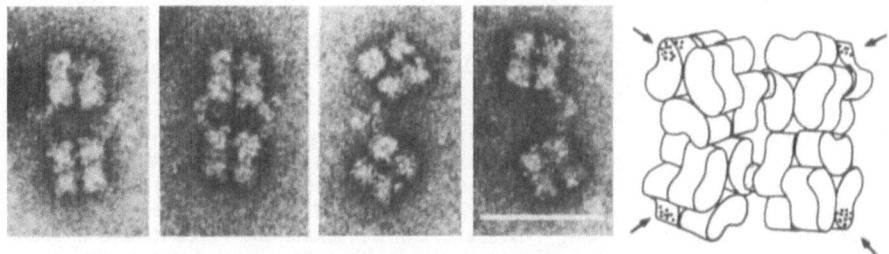

FIG. 10. Intramolecular localization of antigenic determinants in hemocyanin subunits by immunolabelling with monoclonal antibodies.

ACKNOWLEDGMENTS

We are strongly undebted to Mrs Solange Compin and Michèle Leclerc for their skillful technical assistance.

REFERENCES

Bijlholt MMC, van Bruggen EFJ, Bonaventura J. (1979) Dissociation and Reassembly of Limulus polyphemus Hemocyanin. Eur. J. Biochem. 95:399-405

Bijlholt MMC, van Heel MG, van Bruggen EFJ (1982) Comparison of 4x6-Meric Hemocyanins from Three Different Arthropods Using Computer Alignement and Correspondence Analysis. J. Mol. Biol. 161:139-153

Billiald P, Sizaret P-Y, Lamy JN, Feldmann R (1983) Use of Crosslinked Limulus polyphemus Hemocyanin for Intramolecular Localization of Subunits by Immunoelectron Microscopy. Life Chemistry Reports, suppl. 1:43-46.

Brenowitz M, Bonaventura C, Bonaventura J, Gianazza E (1981) Subunit Composition of a High Molecular Weight Oligomer: Limulus polyphemus Hemocyanin. Arch. Biochem. Biophys. 210:748-761

Brenowitz M, Bonaventura C, Bonaventura J (1983) Assembly and Calcium-Induced Cooperativity of Limulus IV Hemocyanin: A Model System for Analysis of Structure-Function Relationship in the Absence of Subunit Heterogeneity. Biochemistry 22:4707-4713

Brenowitz M, Bonaventura C, Bonaventura J (1984a) Selfassociation and Oxygen Binding Characteristics of the Isolated Subunits of Limulus polyphemus Hemocyanin. Arch. Biochem. Biophys. 230:238-249

Brenowitz M, Bonaventura C, Bonaventura J (1984b) Comparison of the Physical and Functional Properties of the 48-Subunits Native Molecule and the 24- and 12-Subunit Dissociation Intermediates of Limulus polyphemus Hemocyanin. Biochemistry 23:879-888

Ellerton HD, Ellerton NF, Robinson HA (1983) Hemocyanin. A Current Perspective. Progr. Biophys. Molec. Biol. 41:143-248

Fearon ER, Love WE, Magnus KA, Lamy J, Lamy J (1983) Crystals of Subunits 2 and 4 of Hemocyanin from the Tunisian Scorpion, Androctonus australis. Life Chemistry Reports, suppl. 1: 65-68

Gaykema WPJ, Hol WGJ, Vereijken JM, Soeter NM, Bak HJ, Beintema JJ (1984) 3.2 Å Structure of the Copper Containing, Oxygen Carrying, Protein Panulirus interruptus Hemocyanin. Nature 309: 23-29

Jollès J, Jollès P, Lamy J, Lamy J (1979) Structural Characterization of Seven Different Subunits in Androctonus australis Haemocyanin. FEBS Letters 106:289-291

Lamy J, Lamy J, Sizaret P-Y, Maillet M, Weill J (1977) Ultrastructure of 16S Substances Obtained by Reassociation Using Different Combinations of Three Isolated Subunits of Scorpion Hemocyanin (Androctonus australis garzonii). J. Mol. Biol. 118:869-875

Lamy J, Lamy J, Weill J (1979a) Arthropod Hemocyanin Structure: Isolation of Eight Subunits in the Scorpion. Arch. Biochem. Biophys. 193:140-149

Lamy J, Lamy J, Weill J, Bonaventura J, Bonaventura C, Brenowitz M (1979b) Immunological Correlates Between the Multiple Hemocyanin Subunits of Limulus polyphemus and Tachypleus tridentatus. Arch. Biochem. Biophys. 196:324-339

Lamy J, Lamy J, Bonaventura J, Bonaventura C (1980) Structure, Function and Assembly in the Hemocyanin System of the Scorpion Androctonus australis. Biochemistry 19:3033-3039

Lamy J, Lamy J, Sizaret P-Y, Weill J (1981a) Quaternary Structure of Androctonus australis Hemocyanin. In: Lamy J and Lamy J (eds) Invertebrate Oxygen Binding Proteins: Structure, Active Site, and Function, Dekker, New York, p 425-443

Lamy J, Bijlholt MMC, Sizaret P-Y, Lamy J, van Bruggen EFJ (1981b) Quaternary Structure of Scorpion (Androctonus australis) Hemocyanin: Localization of Subunits with Immunological Methods and Electron Microscopy. Biochemistry 20:1849-1856

Lamy J, Sizaret P-Y, Frank J, Verschoor A, Feldmann R, Bonaventura J (1982) Architecture of Limulus polyphemus Hemocyanin. Biochemistry 21: 6825-6833

Lamy JN, Lamy J, Sizaret P-Y, Billiald P, Jollès P, Jollès J, Feldmann RJ, Bonaventura J (1983a) The Quaternary Structure of Limulus polyphemus Hemocyanin. Biochemistry 22: 5573-5583

Lamy J, Compin S, Lamy JN (1983b) Immunological Correlates Between Multiple Isolated Subunits of Androctonus australis and Limulus polyphemus Hemocyanins: An Evolutionary Approach. Arch. Biochem. Biophys. 223:584-603

Linzen B (1983) Subunit Heterogeneity in Arthropodan Hemocyanins. Life Chemistry Reports, suppl. 1:27-38

Magnus KA, Love WE (1983) Three-Dimensional Structure of the Limulus II Hemocyanin Subunit at 5.5 Å Resolution. Life Chemistry Reports suppl. 1: 61-64

Markl J, Savel A, Linzen B (1981a) Subunit Composition of Dissociation Intermediates and its Bearing on Quaternary Structure of Eurypelma Hemocyanin. Hoppe-Seyler's Z. Physiol. Chem. 362: 1255-1262

Markl J, Kempter B, Linzen B, Bijlholt MMC, van Bruggen EFJ (1981b) Subunit Topography and a Model of Quaternary Structure of Eurypelma Hemocyanin. Hoppe Seyler's Z. Physiol. Chem. 362:1631-1641

Markl J, Bonaventura C, Bonaventura J (1981c) Kinetics of Oxygen Dissociation from Individual Subunits of Eurypelma and Cupiennius Hemocyanin. Hoppe Seyler's Z. Physiol. Chem. 363:429-437

Markl J, Decker H, Linzen B, Schutter WG, van Bruggen EFJ (1982) The Role of the Individual Subunits in the Assembly of Eurypelma Hemocyanin. Hoppe Seyler's Z. Physiol. Chem. 363:73-87

Schartau WG, Eyerle F, Reisinger P, Geisert H, Starz H, Linzen B (1983) Complete Aminoacid Sequence of Subunit from Eurypelma californicum Hemocyanin and Comparison to chain e. Hoppe Seyler's Z. Physiol. Chem. 364:1383-1409

Schneider HJ, Drexel R, Feldmaier G, Linzen B, Lottspeich F, Henschen A (1983) Complete Aminoacid Sequence of subunit e from Eurypelma californicum Hemocyanin. Hoppe Seyler's Z. Physiol. Chem. 364:1357-1381

Sizaret P-Y, Frank J, Lamy J, Weill J, Lamy J (1982) A Refined Quaternary Structure of Androctonus australis Hemocyanin. Eur. J. Biochem. 127:501-506

Stöffler G, Stöffler-Meilicke M (1984) Immunoelectron Microscopy of Ribosomes. Annu. Rev. Biophys. Bioeng. 13:303-330

van Bruggen EFJ, Bijlholt MMC, Schutter WG, Wichertjes T., Bonaventura J., Bonaventura C, Lamy J, Lamy J, Leclerc, M. Schneider HJ, Markl J, Linzen B (1980) The Role of Structurally Diverse Subunits in the Assembly of Three Cheliceratan Hemocyanins. FEBS Letters 116:207-210 van Heel M, Frank J (1981) Use of Multivariate Statistics in Analysing the Images of Biological Macromolecules. Ultramicroscopy 6:187-194

van Heel M, Keegstra W, Schutter WG, van Bruggen EFJ (1983) Arthropod Hemocyanin Structures Studied by Image Analysis. Life Chemistry Reports, suppl. 1:69-73

van Holde KE, Miller KI (1982) Haemocyanins. Quart. Rev. Biophys. 15:1-129

Cephalopod Hemocyanins: Structure and Function

K.E. van HOLDE, K.I. MILLER

I. INTRODUCTION: THE HEMOCYANINS OF MOLLUSCS

Among the mollusca, hemocyanins are known to serve as oxygen transport pigments in the gastropods, cephalopods, and amphineura. No bivalve has been shown to have hemocyanin; instead, a number use hemoglobin for oxygen transport. All known molluscan hemocyanins share some common features of molecular architecture. The individual polypeptide chains are large (300,000-400,000 daltons) multi-domain structures. Each domain, of which there may be 6-8, is about 50,000 daltons in mass, and carries one oxygen binding site. Under physiological conditions of pH and ionic milieu, these polypeptide chains are associated into much larger structures. Two levels of quaternary organization are commonly observed:

(a) A decamer of polypeptide chains, with sedimentation coefficient about 50-60S and a molecular weight between 3.5×10^6 daltons and 4.0×10^6 daltons.

(b) An icosamer, with sedimentation coefficient of about 100S.

The decamers are arranged as hollow cylinders, with 5 or 10-fold symmetry about the cylinder axis. They usually measure about 300 Å in diameter and 150 Å high. The icosamers are formed by end-to-end association of two decamers. Under some circumstances, higher levels of organization are observed for some molluscan hemocyanins. By virtue of their multi-subunit structure, molluscan hemocyanins are capable of cooperativity in oxygen binding and most, in fact, do show this behavior.

For further details, and a general survey of molluscan hemocyanin structure and function, the reader is referred to either of two recent reviews (Ellerton et al., 1983; van Holde and Miller, 1982). We turn to consideration of the special features of cephalopod hemocyanins.

II. CEPHALOPOD HEMOCYANINS: MOLECULAR STRUCTURE

Although the hemocyanins of a number of cephalopods have been examined by biophysical techniques such as sedimentation (see Ellerton et al., 1983; van Holde and Miller, 1982), there are only a few cases in which sedimentation data have been extrapolated to zero concentration, or in which molecular weights have been determined. These data are summarized in Table 1. Most of the cephalopod molecular weight values in this table are from sedimentation equilibrium experiments; the exceptions are Octopus sloanii pacificus (sedimentation + diffusion) and Sepia officinalis (SDS gels for the monomer, low angle X-ray scattering for the decamer). Perusal of this table yields a number of general conclusions and a few problems.

(a) The decamer, which is invariably the form predominant in the hemolymph, has in all cases a sedimentation coefficient between 51 and 59S, and a molecular weight between 3.5×10^6 and 4×10^6 daltons. These values are consistent with the kinds of structure seen in the electron micrographs, although the sedimentation coefficient of the Octopus decamer seems a bit low, perhaps corresponding to a looser assembly.

(b) In a number of cases, what appears to be a dimeric structure, with a sedimentation coefficient of about 20S, is observed under special solution conditions.

(c) The monomers, obtained by increasing the pH and/or removing divalent cations, seem invariably to have a sedimentation coefficient of about 11S. Most of the molecular weight values obtained for these subunits lie between 3.3×10^5 daltons and 4.2×10^5 daltons. The exceptional result is that obtained by Salvato et al. (1979), who used a number of techniques, including sedimentation equilibrium. All of their results cluster about a value of 2.6×10^5. It is not clear why these values are so much lower than others for cephalopod hemocyanin subunits, but two points should be noted: (i) Some of the experiments were performed under strongly denaturing conditions (i.e. SDS or 6 M GuHCl), and there is evidence that molluscan hemocyanins may contain internal proteolytic cleavages which are only revealed under such circumstances. (ii) In the experiments performed in 3 M urea (e.g. sedimentation equilibrium), no correction was made for preferential binding in the mixed solvent. It is our conclusion that monomer molecular weights of the order of $3.5\text{-}4 \times 10^5$ are generally representative of cephalopod hemocyanins.

Since the mass per oxygen binding site, or per domain, appears to be about the same in cephalopod and gastropod hemocyanins ($\approx 50,000$ daltons), we conclude that cephalopod hemocyanins must contain 7-8 domains. The only case in which a definite result has been established is the Sepia hemocyanin. Bosman et al. (1982) have shown by SDS gel electrophoresis that the whole subunit has a mass of about 4.2×10^5 daltons. By controlled proteolysis, they have been able to demonstrate that the cleavage pattern is consistent with 8 domains. It will be of considerable interest to see if the same pattern will obtain for other cephalopod hemocyanins as well.

TABLE 1. Molecular Parameters of Hemocyanins

Organism	Monomer		Dimer		Decamer		Icosamer		References
	$S^0_{20,w}$	M	$S^0_{20,w}$	M	$S^0_{20,w}$	M	$S^0_{20,w}$	M	
CEPHALOPODS:									
Loligo pealei	11.1	3.85	≈19	7.7	58.7	37.5	-	-	van Holde and Cohen, 1964b
Nautilus pompilius	11.5	3.3	-	-	57.9	35	-	-	Bonaventura et al., 1981
Octopus dofleini	11.1	3.6	≈20	ND	51.0	36	-	-	Miller and van Holde, 1982
Octopus vulgaris	11.3	2.6	ND	ND	+	ND	-	-	Salvato et al., 1979
Ommatostrephes sloanii pacificus	ND	ND	19.5	6.1	+	ND	-	-	Omura et al., 1961
Sepia officinalis	+	4.2	+	ND	+	40	-	-	Bosman et al., 1982 / Pilz et al., 1974
GASTROPOD:									
Helix pomatia (α)	11.2	3.65	19.6	7.6	64.1	42.5	104.3	87	Berger et al., 1976 / Siezen and van Bruggen, 1974 / Konings et al., 1969
Helix pomatia (α)	ND	3.72	ND	ND	ND	37.6	ND	75.5	Herskovits and Russell, 1984

Units: Sedimentation coefficients are in Svedbergs, M in units of 10^5 g/mol.

Symbols: ND = not determined; + = component present, but $S^0_{20,w}$ not measured; - = component not observed.

89

In comparing these results with those of other molluscan hemocyanins, we may note:

(a) Unlike the gastropod and amphineuran hemocyanins, those of the cephalopods never form icosameric (or higher) structures. We have, for example, carried out sedimentation velocity experiments with whole Nautilus hemolymph, and even at this very high concentration ($\simeq 100$ mg/ml) find no evidence for A 100S icosameric component. Small amounts of larger components have been found upon reassociation of subunits (i.e. Bonaventura et al., 1981), but these are clearly artifactual. In contrast, the 100S icosamer is the predominant component in the hemolymph of most gastropods and, in some instances (the Opistho-branchia, in particular), even larger structures are found. Clearly, there are differences in the cephalopod hemocyanin structure which prevent such aggregation.

b) It has generally been considered that the molecular weights of cephalopods hemocyanins are smaller than those of the corresponding gastropod structures. As is shown in Table 10 of van Holde and Miller (1982), most reported values for the icosamers of gastropods are of the order of 9×10^6 daltons. This would correspond to a decamer weight of 4.5×10^6 daltons, in fair agreement with the few values for gastropod decamers that are available. To take a particularly well studied example, values of 8.7×10^6 and 4.25×10^6 have been reported for the Helix pomatia icosamer and decamer, respectively (see Table 1, Berger et al., 1976 and Konings et al., 1969). However, values of 3.6×10^5 daltons (Brouwer et al., 1976) and 3.65×10^5 daltons (Siezen and van Bruggen, 1974) have been obtained for the monomer weight of this protein. These values are in the same range as those found for the cephalopod hemocyanins. Such low values for the monomer weights, combined with the high values for the molecular weights of the 60S and 100S components are, in fact, inconsistent with the postulate that the latter are decamers and icosamer, respectively. In fact, Quitter et al. (1978) found values for the molecular weights of the aggregation states of Busycon hemocyanin which led them to conclude that the 60S and 100S components were 15-mers and 30-mers, respectively. But in most cases the inconsistencies have simply been ignored, with the presumption that the monomer weights were simply too low, for unknown reasons.

To complicate matters, a recent report by Herskovits and Russell (1984) gives quite different values for the molecular weights of the Helix α-hemocyanin. They find M = 7.55 ($\pm$ 0.5) $\times 10^6$ for the icosamer, 3.76 ($\pm$ 0.12) $\times 10^5$ for the decamer, and 3.72 ($\pm$ 0.042) for the monomer (Table 1). These results resemble very much those found for cephalopod hemocyanins, but are in marked desagreement with all earlier data for the gastropod proteins! We are thus left with a dilemma: Are Herskovits and Russell correct (and all earlier studies of gastropod hemocyanins wrong) ? -- in which case, gastropod and cephalopod hemocyanins should be very similar, or -- Are these new data in error ? In the latter case, the peculiar ratios of monomer to multimer molecular weights in gastropod hemocyanins remain to be explained. In favor of higher gastropod weights is the fact that the sedimentation coefficients of gastropod decamers (uniformly about 64S) are substantially larger than those of the cephalopods (see van Holde and Miller, 1982). But it is conceivable that a small tendency toward dimerization has prejudiced both the sedimentation data and the apparent molecular weights of gastropod hemocyanins toward higher values.

In consequence of these uncertainties, we cannot, at the present time, decide whether gastropod and cephalopod monomers and decamers are or are not equivalent in structure.

III. ASSOCIATION-DISSOCIATION EQUILIBRIA OF CEPHALOPOD HEMOCYANINS

The quaternary structures of molluscan hemocyanins are presumably assembled in vivo from monomer chains. Therefore, one might expect that the dissociation of these assemblages would be a reversible process. Dissociation of cephalopod hemocyanins depends upon two factors. First, divalent cations must be removed; complexing with EDTA is effective. Secondly, the pH must be raised above a critical value, about 7.6 for Loligo, 8.5 for Nautilus, and 7.0 for Octopus hemocyanin. Partial dissociation may be accompanied by the appearance of a dimeric species; with Loligo and Octopus hemocyanins it is, with Nautilus it is not. Reassociation can be achieved above the critical pH by reincorporating divalent cations. In the cases of Loligo and Nautilus hemocyanins, it has not been possible to achieve complete reassociation (Bonaventura et al., 1981; van Holde and Cohen, 1964a). Furthermore, many molluscan hemocyanins exhibit a peculiar behavior that indicates that the association-dissociation equilibrium is a complicated one. They often do not obey the mass action law; that is, changes in total hemocyanin concentration do not produce the expected shifts in the association-dissociation equilibrium (see Engelborghs and Lontie, 1973; Konings et al., 1969; Wood and Peacocke, 1973, for example). Various explanations have been put forward for this curious behavior, but all involve the assumption of some kind of microheterogeneity in the quaternary structures. If there is more than one kind of subunit, and different combinations have differing stabilities, the problems encountered in reassociation could also be explained.

For these reasons, we were very surprised to find that the hemocyanin of Octopus dofleini is capable of quantitative reassociation from 11S subunits by the simple addition of appropriate cations (see figure 1). We have investigated this behavior in some detail, and find that a monomer-decamer equilibrium exists in the neighborhood of pH 8.0. Under these conditions, no detectable quantity of the dimer is present, and the reassociation reaction is sufficiently slow that sedimentation velocity studies can be used to measure the amounts of monomer and decamer. We find the equilibrium obeys the mass action law (Fig. 2) and is extremely sensitive to calcium and magnesium ion concentrations (Fig. 3). A divalent cation is not essential, however, for sodium ions will also promote association, at sufficiently high concentrations (Fig. 3).

It is clearly cations that are involves for all of the solutions described in figure 3 have the same anion (Cl^-), but at radically different concentrations. We believe that cations are effectual by screening carboxylate ions which will otherwise inhibit association. This concept is supported by the fact that below pH 7 the decameric structure is stable even in the absence of added cations.

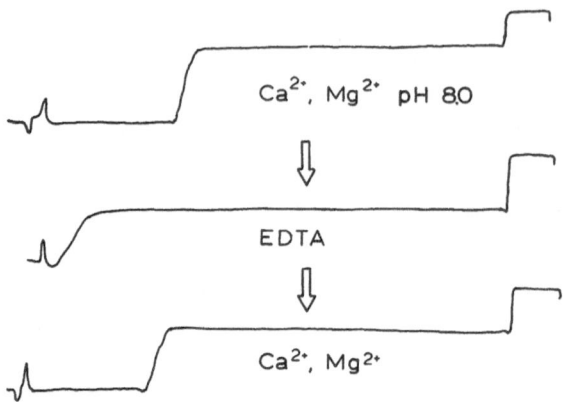

FIG. 1. Demonstrating that <u>Octopus</u> <u>dofleini</u> hemocyanin is capable of quantitative reassociation. In the top panel is shown a scanner trace of sedimentation of the decameric hemocyanin, at pH 8.0, in the presence of 50 mM Mg^{2+}, 10 mM Ca^{2+}. The center panel shows a scan of the 11S monomer obtained by dialysis of the above solution against 10 mM EDTA. In the bottom panel is shown the result of dialyzing the solution back into 50 mM Mg^{2+}, 10 mM Ca^{2+}. All of the hemocyanin has reassociated; no remaining 11 S material is detectable.

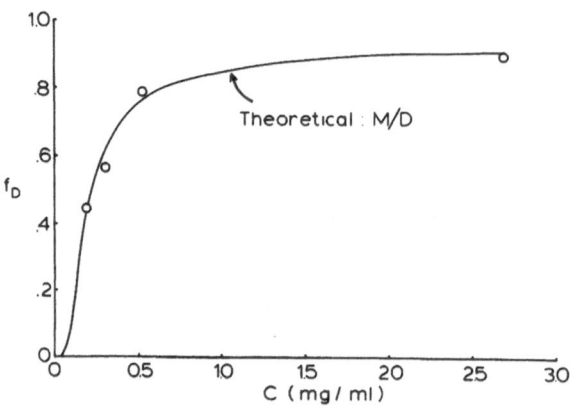

FIG. 2. The dissociation-association equilibrium of <u>Octopus</u> <u>dofleini</u> hemocyanin obeys the mass action law. A concentrated solution was prepared in 8 mM Mg^{2+}, pH 8.0. This was then diluted to three lower concentrations, and all solutions allowed to equilibrate at 20 °C. The fraction of decamer in each (f_D) was then determined by sedimentation analysis. The solid line is a theoretical curve for a monomer $\rightleftharpoons$ decamer equilibrium.

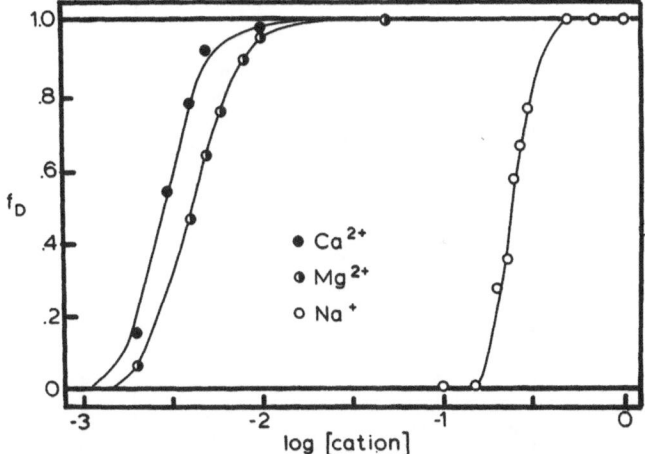

FIG. 3. The effects of Ca^{2+}, Mg^{2+} and Na^+ on the monomer-decamer equilibrium of Octopus hemocyanin at pH 8.0. The fraction decamer in each case was determined by sedimentation analysis.

IV. OXYGEN BINDING BY CEPHALOPOD HEMOCYANINS

The primarily physiological function of hemocyanins is, of course, oxygen transport. Given the overall similarities of the cephalopod hemocyanins in subunit structure and quaternary structure, the differences observed in oxygen binding behavior are remarkable. Table 2 summarizes data on intact (decameric) hemocyanin molecules, under conditions approaching the physiological environment. As can be seen, the O_2 affinity varies greatly; Nautilus hemocyanin has a quite high affinity, whereas Loligo has one of the lowest values ever observed. Whereas all known cephalopod hemocyanins have normal Bohr effects, the magnitude ranges from very small in Sepia and Nautilus to an extremely large effect in Octopus. All seem to exhibit some degree of cooperativity in binding, as indicated by the Hill coefficients, but these values are considerably smaller than would be possible considering the number of subunits in the protein. Each of these decamers must contain between 60-80 oxygen binding sites, but the Hill coefficients have never been observed to exceed 4.

Since the maximum Hill coefficient is actually less than the number of oxygen binding sites in each subunit, it might be thought that assembly of these subunits into a quaternary structure would not be necessary for allosteric behavior. However, in three cases studied to date (Loligo, Nautilus, Octopus) it is found that dissociation invariably leads to the loss of cooperativity in oxygen binding. Thus, it appears that constraints imposed by the decameric structure are essential to the linkage of sites. In accord with this idea is the observation of DePhilipps et al. (1969) that partially oxygenated Loligo hemocyanin showed a strong tendency to dissociate; it is as if the oxy- and deoxy-conformations of subunits are tendency to dissociate; it is as if the oxy- and deoxyconformations of subunits are different, and do not fit comfortably within the same molecule.

TABLE 2. Oxygen-Binding Parameters of Cephalopod Hemocyanins

Organism	T (°C)	pH	Ionic Conditions	P_{50}	n_m	$\dfrac{d\log P_{50}}{dpH}$	References
Loligo pealei	23	7.4	10 mM Mg^{2+}	150	3.9	-1.0	DePhilipps et al., 1969
Nautilus pompilius	20	7.4	50 mM Mg^{2+}, 10 mM Ca^{2+}	8	2.0	-0.2	Bonaventura et al., 1981
Octopus dofleini	20	7.4	50 mM Mg^{2+}, 10 mM Ca^{2+}	16	3.2	-1.7	K. Miller (unpublished)
Sepia officinalis	20	7.4	1 mM Tris; concentrated solution	26	nd	-0.1	Wolf and Decleir, 1981

Symbols: P_{50} = oxygen pressure at half-saturation; n_m = maximum slope of Hill plot; $d\log P_{50}/dpH$ = Bohr coefficient.

The Bohr effect in cephalopod hemocyanins also appears to be complex. Recent studies by Miller (unpublished) show that while individual binding curves can be fitted by the model of Monod et al. (1965), the binding affinities in both the "tense" and "relaxed" states are functions of pH. This contrasts strongly with the behavior of some arthropod hemocyanins, such as Penaeus (Brouwer et al., 1978) and Callianassa (Arisaka and van Holde, 1979; Miller and van Holde, 1974) for which the Bohr effect can be explained entirely in terms of pH-dependence of the allosteric parameter L.

ACKNOWLEDGMENTS

This research was supported in part by a grant from the National Science Foundation.

REFERENCE

Arisaka F, van Holde KE (1979) Allosteric properties and association equilibria of hemocyanin from Callianassa californiensis. J. Mol. Biol. 134: 41-73

Berger J, Pilz I, Witters R, Lontie R (1976) Röntgenklein-winkel-und sedimentations studien am α-Hämocyanin Helix pomatia (halbe moleküle) in glycerine-und saccharose lösungen. Z. Naturf. 31: 238-244.

Bonaventura C, Bonaventura J, Miller KI, van Holde KE (1981) Hemocyanin of the chambered nautilus: Structure-function relationships. Arch. Biochem. Biophys. 211: 589-598

Bosman F, Gielens C, Préaux G, Lontie R (1982) Limited proteolysis of the hemocyanin of Sepia officinalis. Arch. Int. Physiol. Biochim. 90: 1384

Brouwer M, Wolters M, van Bruggen EFJ (1976) Proteolytic fragmentation of Helix pomatia α-hemocyanin. Structural domains in the polypeptide chain. Biochemistry 15: 2618-2623

Brouwer M, Bonaventura C, Bonaventura J (1978) Analysis of the effect of 3 different allosteric ligands on oxygen binding by hemocyanin of the shrimp, Penaeus setiferus. Biochemistry 17: 2148-2154

DePhilipps HA, Nickerson KN, Johnson M, van Holde KE (1969) Physical studies of hemocyanins. IV. Oxygen linked dissociation of Loligo pealei hemocyanin. Biochemistry 8: 3665-3672

Ellerton HD, Ellerton NF, Robinson HA (1983) Hemocyanin: A current perspective. Progr. Biophys. Mol. Biol. 41: 143-248

Engelborghs Y, Lontie R (1973) Dissociation of Helix pomatia haemocyanin under the influence of alkali salts. J. Mol. Biol. 77:577-587

Herskovits TT, Russell MW (1984) Light scattering investigation of the subunit structure and dissociation of Helix pomatia hemocyanin. Effects of salts and ureas. Biochemistry 23: 2812-2819

Konings WN, Siezen RJ, Gruber JM (1969) Structure and properties of hemocyanins. XI. Association-dissociation behavior of Helix pomatia hemocyanin. Biochim. Biophys. Acta 194: 376-385

Miller KI, van Holde KE (1974) Oxygen binding by Callianassa californiensis hemocyanin. Biochemistry 13: 1668-1674

Miller KI, van Holde KE (1982) The structure of Octopus dofleini hemocyanin. Comp. Biochem. Physiol. 73B: 1013-1018

Monod J, Wyman J, Changeux JP (1965) On the nature of allosteric transition: A plausible model. J. Mol. Biol. 12: 88-118

Omura T, Fujita T, Yamada F, Yamamoto S (1961) Hemocyanin of Ommatostrephes sloanii pacificus. J. Biochem. 50: 400-404

Pilz I, Engelborghs Y, Witters R, Lontie R (1974) Studies by X-ray small-angle scattering of the quaternary structure in solution of halves and tenths of Helix pomatia haemocyanin and of Sepia officinalis haemocyanin. Eur. J. Biochem. 42: 195-202

Quitter S, Watts LA, Crosby C, Roxby R (1978) Molecular weights of aggregation states of Busycon hemocyanin. J. Biol. Chem. 253: 525-530

Salvato B, Ghiretti-Magaldi A, Ghiretti F (1979) Hemocyanin of Octopus vulgaris. Molecular weight of the minimal functional subunit in 3 M urea. Biochemistry 18: 2731-2736

Siezen RJ, van Bruggen EFJ (1974) Structure and properties of hemocyanin. XII. Electron microscopy of dissociation products of Helix pomatia α-hemocyanin: Quaternary structure. J. Mol. Biol. 96: 77-80

van Holde KE, Cohen LB (1964a) The dissociation and reassociation of Loligo pealei hemocyanin. Brookhaven Symp. Biol. 17;184-193

van Holde KE, Cohen LB (1964b) Physical studies of hemocyanins. I. Characterization and subunit structure of Loligo paelei hemocyanin. Biochemistry 13: 1803-1808

van Holde KE, Miller KI (1982) Haemocyanins. Quarterly Rev. Biophys. 15: 1-129

Wolf G, Decleir W (1981) A study of hemocyanin in Sepia officinalis: Functional properties of the adult molecule. In: Lamy J and Lamy J (eds) Invertebrate Oxygen Binding Proteins. Marcel Dekker, New York, pp. 749-754

Wood EJ, Peacocke AR (1973) Murex trunculus haemocyanin. I. Physical properties and pH-induced dissociation. Eur. J. Biochem. 35: 410-420

Molecular and Cellular Adaptations of Fish Hemoglobin-Oxygen Affinity to Environmental Changes

D.A. POWERS

I. INTRODUCTION

Comparative biochemists and physiologists have found that the blood-oxygen affinities of various fish species are compatible with the physical and chemical parameters of their environments. For example, fish that live in low oxygen environments have high oxygen affinities while those that live in high oxygen environments have lower oxygen affinities. Moreover, fish that live in environments where physical parameters periodically change, have the necessary molecular machinery required for adaptation. This machinery includes species specific hemoglobins and/or the regulation of various modifier ligands (e.g., organic phosphates, HCO_3, CO_2, Cl, H^+, etc.). The intraerythrocyte concentrations of these ligands influence oxygen binding and are directly or indirectly effected by environmental parameters (e.g. temperature). It is, therefore, appropriate to review some aspects of hemoglobin as it relates to these interactions. Once an overview of these linked functions has been established, we shall address some structural and functional aspects of multiple hemoglobins, then turn our attention to the molecular mechanism of adapting to environmental oxygen and temperature to illustrate these phenomena.

II. MOLECULAR MECHANISMS FOR MAINTAINING RESPIRATORY HOMEOSTASIS

The primary function of hemoglobin is to carry oxygen from the "organism-environment interface" (e.g. the gills, lungs, etc.) to the respiring tissues. This physiological function depends on the hemoglobins' ability to form a reversible complex between oxygen and the ferrous iron in the hematoporphyrin. Under constant physical conditions and in the absence of modifying molecules, the affinity of the hemoglobin for oxygen depends primarily on the hydrophobic nature of the heme pocket and the roles of specific amino acid residues that either directly or indirectly affect hemoglobin-oxygen ($Hb-O_2$) affinity. Since these amino acids are coded by the globin genes, the intrinsic $Hb-O_2$ affinity is genetically determined. This level of adaptation is an evolutionary strategy that involves the genetically determined intrinsic properties of the hemoglobin.

The interactions between heme sites in the binding of oxygen (i.e., homotropic interactions) are under genetic control, but considerable flexibility is provided by the binding of modifier ligands (i.e., heterotropic interactions). While the affinity constants for these ligands are genetically determined, plasticity is provided by regulating intracellular concentrations of the allosteric modifiers. The mathematical concept that describes the interactions between a macromolecule and its various ligands was formalized by Wyman (1948; 1964).

A. An Overview of Linked Functions

1. Variable Oxygen and Organic Phosphate

When a macromolecule (M) like hemoglobin binds four ligands (X) (like oxygen) and one molecule of a second ligand (D) like an organic phosphate, the simultaneous equilibria can be depicted following linkage scheme:

$$
\begin{array}{ccccccccc}
 & & & & {}^{4x}K & & & & \\
 & {}^{x}k & & {}^{2x}k & & {}^{3x}k & & {}^{4x}k & \\
M & \longleftrightarrow & MX & \longleftrightarrow & MX_2 & \longleftrightarrow & MX_3 & \longleftrightarrow & MX_4 \\
{}^{D}K \updownarrow & & {}^{D}K_x \updownarrow & & {}^{D}K_{2x} \updownarrow & & {}^{D}K_{3x} \updownarrow & & {}^{D}K_{4x} \updownarrow \\
MD & \longleftrightarrow & MXD & \longleftrightarrow & MX_2D & \longleftrightarrow & MX_3D & \longleftrightarrow & MX_4D \\
 & {}^{X}k_D & & {}^{2x}k_D & & {}^{3x}k_D & & {}^{4x}k_D & \\
 & & & & {}^{4x}K_D & & & & \\
\end{array}
$$

(Scheme 1)

Where the k's are the microconstants for individual oxygenation steps, and the K's are macroconstants. The <u>presuperscripts</u> on all constants indicate the number and type of ligands <u>being bound</u>; while the subscripts indicate the number and type of ligands <u>already bound</u> to the macromolecules.

If one assumes that the temperature, pH, etc., are held constant and that the hemoglobin does not dissociate or polymerize, this system can be described by 9 of the 13 equilibrium constants. The Adair constants for oxygen binding for scheme 1 would be:

$$
{}^{ix}K = \frac{\left[MX_i\right]}{\left[M\right]\left[X\right]^i} \qquad \text{and} \qquad {}^{ix}K_D = \frac{\left[MDX_i\right]}{\left[MD\right]\left[X\right]^i} \tag{Eq. 1}
$$

where i = 1, 2, 3, 4

while the organic phosphate association constants can be described by:

$$^{D}K_{jx} = \frac{\left[MX_j\right]}{\left[MX_j\right]\left[D\right]}$$

(Eq. 2)

where j = 0, 1, 2, 3, 4 and $^{D}K_{jx} = ^{D}K$ when j = 0.

Ackers (1979) has suggested that the nine most appropriate exprimentally determinable constants might be those described in Eq. 1 plus ^{D}K and $^{D}K_{4x}$ (Eq. 2 where j = 0 and 4, respectively).

At a defined set of conditions (i.e., temperature, pH, ionic strength, etc.) the Adair constants can be determined by oxygen binding experiments in the absence and presence of saturating concentrations of organic phosphate, respectively. We have shown that ^{D}K and $^{D}K_{4x}$ (scheme 1) can be determined by organic phosphate binding studies in the absence and presence of oxygen, respectively (Powers et al., 1981). The use of Ackers' (1979) exact theory allows the elucidation of the constants for intermediate oxygen concentrations. Those constants combined with the values determined by oxygen and organic phosphate binding experiments can provide a comprehensive thermodynamic profile of the interactions of hemoglobin and its ligands.

2. Variable Proton Concentration

Scheme 1 is valid only at defined pH. When proton concentration is allowed to vary it affects oxygen binding as well as the binding of organic phosphate. At relatively high pH (e.g., pH 10) there is little or no Bohr or organic phosphate effects on Hb-O_2 affinity. Conversely, at low pH there can be a reversal or inhibition of both the Bohr and organic phosphate effects on Hb-O_2 affinity. Thus, the Bohr and organic phosphate effects are most pronounced over a limited pH range. Assuming that the protons that affect the binding of organic phosphate are between o and z for deoxyhemoglobin and between o and r for oxyhemoglobin, the interactions can be described for hemoglobin by:

(Scheme 2)

and for oxyhemoglobin by:

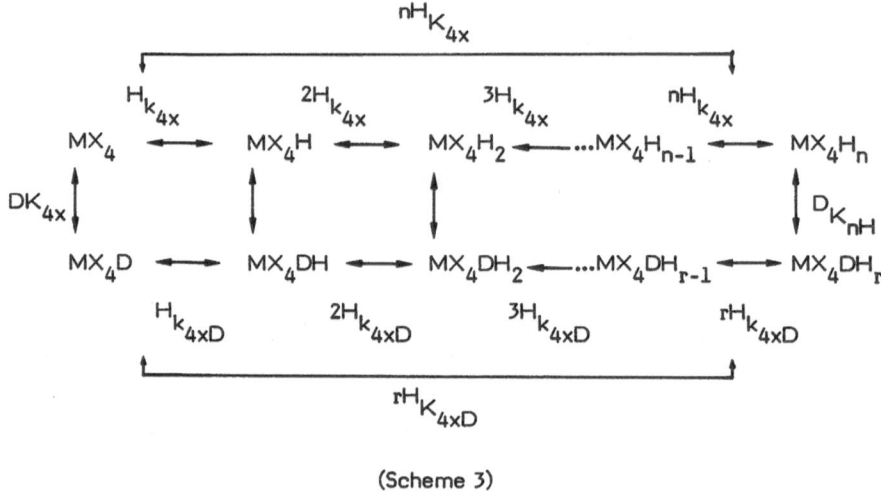

(Scheme 3)

3. Variable Oxygen, Protons and Organic Phosphate

Figure 1 represents the macro-interaction between hemoglobin, oxygen, organic phosphate, and protons. Scheme 2 represents the binding of protons that affect organic phosphate affinity for deoxyhemoglobin (protons o to z).

Scheme 3 similarly represents that portion of figure 1 that deals with oxyhemoglobin for protons o to r.

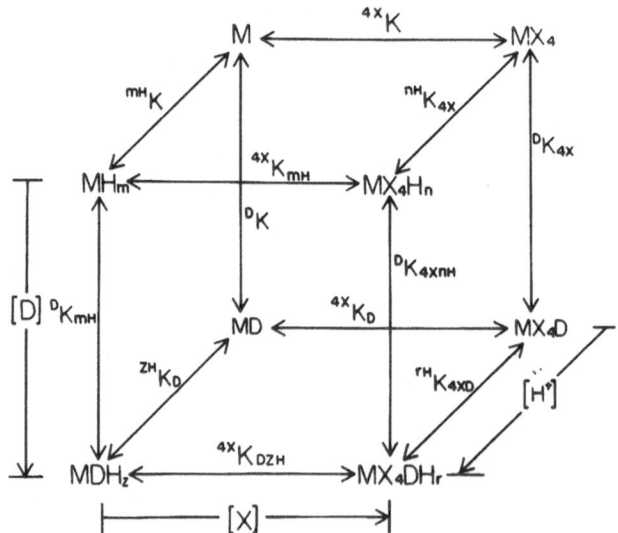

FIG. 1. A generalized schematic of the linkage equilibria which relate the binding of protons (H), organic phosphate (D) and oxygen (X) to hemoglobin tetramers (M). Hemoglobin with oxygen bound is represented with all four sites saturated. The edges of the diagram indicate increasing concentration of appropriate ligand (H, X, and D) by arrows.

An experimentally determinable organic phosphate-hemoglobin binding constant is a composite of the individual affinities for the various molecular species involved.

When the hemoglobin is at the extremes of oxygenation, (see Fig. 1) the apparent binding of organic phosphate to oxy- or deoxyhemoglobin can be measured directly (Powers et al., 1981). The macroconstants for the other faces of figure 1 are obtainable from oxygen binding measurements carried out as a function of pH and organic phosphate concentrations.

For deoxyhemoglobin (M) the apparent binding constant for organic phosphate $(^DK^{app}_{deoxy})$ is defined by:

$$^DK^{app}_{deoxy} = (^DK) \frac{^{zH}Z_{deoxy}}{^{mH}Z_{deoxy}} \qquad \text{(Eq. 4)}$$

where $^{zH}Z_{deoxy}$ and $^{mH}Z_{deoxy}$ are the binding polynomials for deoxyhemoglobin

$$^{mH}Z_{deoxy} = \sum_{i=o}^{m} {}^{iH}k \left[H\right]^i$$

$$^{zH}Z_{deoxy} = \sum_{i=o}^{z} {}^{iH}k_D \left[H\right]^i$$

where

$$^{iH}k \equiv 1 \text{ and } {}^{iH}k_D \equiv 1 \text{ when i = o}$$

If we define the macroconstants for putting on j protons in the absence of organic phosphate as:

$$^{jH}K = \prod_{i=o}^{j} {}^{iH}k \quad \text{where} \quad {}^{iH}k \equiv 1 \quad \text{when i = o}$$

and putting on j protons on deoxyhemoglobin in the presence of organic phosphate as:

$$^{jH}K_D = \prod_{i=o}^{j} {}^{iH}k_D \quad \text{where} \quad {}^{iH}k_D \equiv 1 \quad \text{when i = o}$$

After substituting the various macroconstants, the apparent constant for organic phosphate binding to deoxyhemoglobin would be:

$$^DK^{app}_{deoxy} = \frac{^DK\left(1 + {}^{H}K_D\left[H\right] + {}^{2H}K_D\left[H\right]^2 + \dots {}^{zH}K_D\left[H\right]^z\right)}{\left(1 + {}^{H}K\left[H\right] + {}^{2H}K\left[H\right]^2 + {}^{3H}K\left[H\right]^3 + \dots {}^{mH}K\left[H\right]^m\right)} \qquad \text{(Eq. 5)}$$

The macroconstants for oxyhemoglobin are also defined as products of the individual constants:

$$^{JH}K_{4x} = \prod_{i=o}^{j} {}^{iH}k_{4x}$$

$$^{JH}K_{4x} = \prod_{i=o}^{j} {}^{iH}k_{4xD}$$

where $^{iH}K_{4x} \equiv 1$ and $^{iH}K_{4xD} = 1$ when i = o

Therefore, the equation for organic phosphate binding to oxyhemoglobin is:

$$^DK^{app}_{deoxy} = \frac{^DK_{4x}\left(1 + {}^HK_{4xD}\left[H\right] + {}^{2H}K_{4xD}\left[H\right]^2 + \ldots {}^{rH}K_{4xD}\left[H\right]^r\right)}{\left(1 + {}^HK_{4x}\left[H\right] + {}^{2H}K\left[H\right]^2 + \ldots {}^{nH}K_{4x}\left[H\right]^n\right)} \quad \text{(Eq. 6)}$$

The various $^DK^{app}$ values can be determined for oxy- and deoxyhemoglobin at defined pH values by the method of Powers et al. (1981). These data cannot be fitted to equations 5 and 6 easily because there are an infinite number of possible constants. Since we know only a few protons are involved, the data can be fitted, via non-linear least squares analysis to simplified versions of equations 5 or 6 (Hobish and Powers, 1985). Representative data and fitted curves for Carp I deoxyhemoglobin (Greaney et al., 1980a) and human oxy- and deoxyhemoglobin (Hobish and Powers, 1985) can be seen in figure 2.

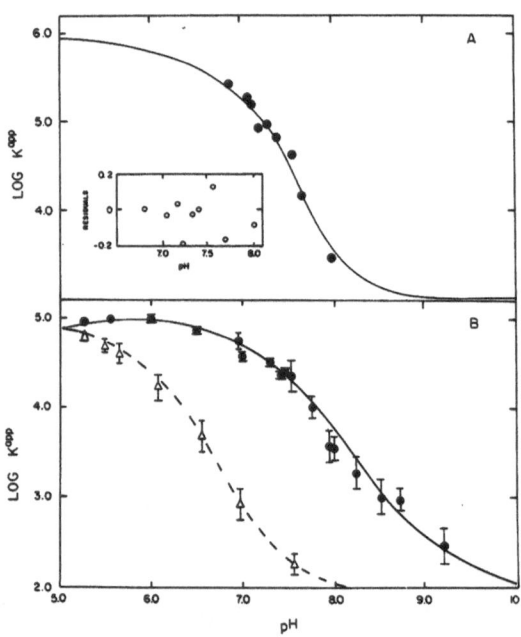

FIG. 2. Representative data obtained by the described method. (A) Plot of the variation of the common logarithm of the experimental association constant (K^{app}) with pH for ATP binding to carp deoxyhemoglobin I at 30 °C. Association constants are reciprocals of dissociation constants, which were determined from the slopes of Scatchard plots. The curve fitting was accomplished by the nonlinear least-squares minimization method alluded to in the text. The inset shows the residuals of the best fit. (B) Plot of the variation of the common logarithm of the experimental association constant (K^{app}) with pH for DPG binding to human hemoglobin A_0 at 21.5 °C. The open symbols represent the values for oxyhemoglobin, and the filled symbols represent data for deoxyhemoglobin. Association constants were obtained by fitting the data via a nonlinear least-squares analysis to a Langmuir isotherm.

These simplified equations (equations 6 and 7) are functionnaly similar to a Hill equation

$$^DK^{app}_{deoxy} = \frac{^DK\left(1 + {}^{ZH}K_D\left[H\right]^z\right)}{\left(1 + {}^{MH}K\left[H\right]^m\right)} \quad \text{(Eq. 7)}$$

$$D_K{}^{app}_{deoxy} = \frac{{}^{D}K_{4x}\left(1 + {}^{rH}K_{4xD}\left[H\right]^r\right)}{\left(1 + {}^{nH}K_{4x}\left[H\right]^n\right)} \qquad \text{(Eq. 8)}$$

III. MULTIPLE HEMOGLOBINS

One strategy for adapting to a changing environment would be to have a number of hemoglobins each with unique functional characteristics. In contrast to the hemoglobins of most mammals and birds, the presence of multiple hemoglobins is common in fishes (reviewed by Riggs, 1970). Closely related groups of fish may differ widely in their number of isohemoglobins (e.g., Riggs, 1970; Bonaventura et al., 1975). In some species the hemoglobin loci are polymorphic (Fyhn and Sullivan, 1974; Hjorth, 1975; Fyhn et al., 1979). Multiple hemoglobins and polymorphic loci (i.e. isohemoglobins and allohemoglobins) may be adaptative responses to a variable environment.

A. Structural Aspects of Multiple Hemoglobins

While the presence of multiple hemoglobins in lower vertebrates is a well-established phenomena, there are discrepancies regarding the actual number, proportion, structural and functional properties of species-specific isohemoglobins (reviewed by Fyhn and Sullivan, 1975; Riggs, 1970). Such differences can arise from thermoacclimatory variation (Houston and Cyr, 1974) developmental changes (Iuchi, 1973; Watt and Riggs, 1975), polymerization (Fyhn and Sullivan, 1975; Reischl, 1976), dissociation (Briehl, 1963; Fyhn and Sullivan, 1975), and susceptibility to autooxidation (Tsuyuki et al., 1965; Yamanaka et al., 1965). In addition, other technical aspects related to erythrocyte preparation can lead to electrophoretic artifacts. One of the most interesting sources of electrophoretic variation arises from genetically unique globin chains that combine to give a series of unique hemoglobin tetramers. Unfortunately, due to numerous sources of uncertainty, an analysis of the subunit structure of multiple fish hemoglobins has only rarely been attempted. In those instances where such analysis has been done, two adaptive strategies have become evident. First, there are multiple hemoglobins that are structurally different by functionally equivalent. Second, there are structurally different hemoglobins, some of which (but not necessarily all) are functionally unique.

1. Hemoglobins from Fundulus heteroclitus

The fish Fundulus heteroclitus has four isohemoglobins, each a tetramer of about 64 kDa (Mied and Powers, 1978). They have different isoelectric points (between pH 5-9) and show very little tetramer-dimer dissociation in the presence of 0.1 M NaCl, 1.0 M NaCl, and 4 M urea.

There are four different globin chains α^a, α^b, β^a and β^b. The HbI is a homotetramer composed of two α^b and two β^b chains, HbIV is a homotetramer consisting of two α^a and two β^a subunits. Isohemoglobins II and III are heterotetramers consisting of all four chains. While the amino acid composition of the chains are significantly different (Mied and Powers, 1978), the end-groups of homologous chains (i.e., α^a: α^b and β^a: β^b) are identical (Table 1). Clearly, all four hemoglobins of this fish species are structurally unique, however, there are no significant differences in their Hb-O_2 affinities (Mied and Powers, 1978; Powers, 1980).

TABLE 1. The End Groups of the Isolated Globin Chains from Fishes

Type of Fish Hb	Model species	Globin chain	NH_2-terminal	COOH terminal
Class I Hemoglobins	Fundulus heteroclitus	β^a	Val-	-Tyr-His
		β^b	Val-	-Tyr-His
		α^a	Ac-Ser	-Tyr-Arg
		α^b	Ac-Ser	-Tyr-Arg
	Cyprinus carpio	β^a	Val-Glu-Trp	
		β^b	Val-Glu-Trp	-Tyr-His
		α	Ac-Ser-	-Tyr-Arg
Class II Hemoglobins	Catostomus clarkii	β^a	Val-Glu-Trp	-Tyr-His
		β^b	Val-Glu-Trp	-Tyr-His
		β^c	Val-Glu-Trp	-Tyr-Phe
		α^a	Ac-Ser-	-Tyr-Arg
		α^b	Ac-Ser-	-Try-Arg
	Trout IV	β	Val-Asp-Trp	-Tyr-His
		α	Ac-Ser	-Tyr-Arg
	I	β	Val-Glu-Trp	-Tyr-Phe
		α	Ac-Ser	-Tyr-Arg

2. Hemoglobins from Carp

Hemolysates of red cells from carp contain three hemoglobin components (Gillen and Riggs, 1972). The major component HbI has been one of the better studied fish hemoglobin both structurally and functionally. Each hemoglobin is composed of a tetramer of two α and two β chains. The complete amino acid sequence of the α chain was determined by Hilse and Braunitzer (1968) and the two β chain sequences were published by Grujic-Injac, Braunitzer and Stangl (1980). While the other two hemoglobins (HbII and HbIII) have not been completely analyzed, there appears to be at least one additional β chain (Gillen and Riggs, 1972) and perhaps a minor α chain (Hilse and Braunitzer, 1968). The end groups of the α and β chains (Table 1) are similar to those of $\underline{F.}$ heteroclitus hemoglobins and the anodal hemoglobins of Catostomus and Trout HbIV (Table 1). There are no apparent remarkable differences in the Hb-O_2 affinities of these hemoglobins.

3. Hemoglobins of Catostomus clarkii

The fish Catostomus clarkii has 12 isohemoglobins comprised of at least two α chains (α^a and α^b) and four β chains (β^a, β^b, β^c and β^d). The complete amino acid sequence of one α chain and partial sequences of two β chains has been studied (Powers and Edmundson, 1972b). The end groups of the α end and β chains for the electrophoretically anodal hemoglobins are similar to the hemoglobins of $\underline{F.}$ heteroclitus and Carp, however, the β chains of the cathodal hemoglobins have a COOH-terminal Phe instead of His (Table 1). This important amino acid difference, (Powers, 1972), was later also found to be shared by trout HbI (Table 1, Barra et al., 1973). This COOH-terminal residue is important in the function of the hemoglobin in relation to physiologically significant changes in pH (see Powers, 1972 and later discussion below).

4. Hemoglobins of Trout

The trout hemoglobins are a particularly good model to study structural and functional differences. They have many of the interesting properties of the C. clarkii hemoglobins, but have 4 rather than 12 hemoglobins. The four major hemoglobins are numbered I, II, III and IV, respectively. The trout IV hemoglobin, like the anodal hemoglobins of C. clarkii, has a Bohr effect and responds to heterotrophic ligands while I and II do not. There are at least two α and two β chains. The β chain residues of HbI are believed to be involved in both the Bohr effect and organic phosphate binding sites have been replaced by neutral residues (Perutz, 1984). The sequences of both HbI and HbIV have been completed (Barra et al., 1983; Bossa et al., 1976, 1978). This elegant structural work, coupled with functional studies (reviewed by Brunori, 1975) and the tertiary structural implications elucidated by Perutz (1984), have provided valuable insight concerning the physiological and ecological adaptations suggested by my work on Catostomus hemoglobins (Powers, 1972; Powers and Edmundson, 1972a, 1972b).

B. Subunit Contacts

The subunit contacts of vertebrate hemoglobins can be divided into two classes: $\alpha_1 \beta_1$ and $\alpha_1 \beta_2$. Most of the $\alpha_1 \beta_2$ contacts have been conserved over evolutionary time.

About half of the $\alpha_1\beta_2$ contact residues have remained identical and most of the other residues are the result of very conservative isopolar interchanges. The $\alpha_1\beta_1$ contacts are more extensive in number than the $\alpha_1\beta_2$ contacts. Variation in the $\alpha_1\beta_1$ contacts are also more prevalent than those of the $\alpha_1\beta_2$. In fact, only 25 to 30% of the $\alpha_1\beta_1$ contact residues are identical across divergent species and most of the variable contacts have substitutions that are neither conservative nor isopolar. Interestingly, however, the evolutionarily conserved β chain residues tend to be those complementary to residues that are conserved in the α chains. For example, H5, H9 and H10 of the α chain are identical across species and most of their contacts in the β chain (e.g., B12, B16, C1 and C3) have also been conserved. Thus, evolution appears to have selected a few complementary $\alpha_1\beta_1$ contacts rather than the residues per se, but most of the $\alpha_1\beta_1$ contact residues have not been highly conserved.

C. Heme Contacts

The residues involved in heme contacts have been highly conserved over evolutionary time. Seventy five per cent of the α chain and 50% of the β chain heme contacts are identical across divergent species. In addition, almost all of the variable heme contacts are the result of very conservative isopolar substitutions (e.g., Ser $\longrightarrow$ Thr, Phe $\longrightarrow$ Tyr, Ala $\longrightarrow$ Gly, Leu $\longrightarrow$ Ile, etc.).

While significant changes in primary structure have occurred during evolutionary divergence, the molecular integrity of the subunits and the hemoglobin tetramers has been maintained. An inspection of the homologous $\alpha_1\beta_2$ subunit and heme contacts of evolutionarily divergent vertebrate species suggest natural selection against extreme variability of many, but not all, of these residues. Since maintenance of structural integrity is essentially to efficient function of the hemoglobin (Perutz and Lehmann, 1968; Perutz et al., 1969), the variability of any residue is restricted by its position in the three-dimensional structure. It follows that different residue positions will have different ranges of evolutionary flexibility, depending upon their structural and physiological restrictions. Consequently, the range of amino acid variability can be drastically restricted by specific structural limitations. Some of these limitations seem to be imposed by residues involved in the heme pockets, the $\alpha_1\beta_2$ interface and a few specific $\alpha_1\beta_1$ interface residues. Moreover, it cannot be assumed that residue variability means that those positions are not important. The range of Hb-O$_2$ affinity and responses to various heterotrophic interactions among evolutionary divergent species differ by several orders of magnitude. Thus, some of the variable residues involved in subunit and heme contacts as well as elsewhere could be responsible for species specific functional differences. Of course, some residues may be functionally equivalent, but such equivalence must be proven by appropriate experimental analysis rather than circular reasoning. These conclusions are similar to those stated a decade ago (Powers and Edmundson, 1972b) employing a much more limited data set.

D. Functional Aspects of Multiple Hemoglobins

1. Bohr effect

Like mammals, most fish hemoglobins have Hb-O_2 affinities that are minimized at low pH values and maximized at higher values. This phenomenon (the Bohr effect) is beneficial in the binding of oxygen at the gills and its release at respiring tissues. Some fish have hemoglobins with such a large "Bohr effect" that it cannot be saturated with oxygen at a very low pH, even when exposed to pressures that are several times that of atmospheric oxygen. The phenomenon termed the "Root Effect" (Root, 1931) appears to be particularly important in delivering oxygen to the swim bladder, so that neutral buoyancy can be maintained. In addition, there also appears to be a physiological role in the delivery of oxygen to retinal tissue in some fish (Wittenberg and Wittenberg, 1974). Perutz (1984) has recently reviewed the major amino acids residues responsible for the Bohr and Root effects.

Although a large Bohr effect is beneficial in releasing oxygen at the cellular level, it can suppress oxygen binding at the gills if blood pH drops sufficiently low. When swimming stamina tests are conducted on fish in a closed water tunnel, a fish's blood pH dramatically drops and eventually the fish can no longer swim (DiMichele and Powers, 1982b; Powers, 1980). This reduction in pH is accompanied by an increase in both hematocrit and serum lactate; the latter of which is produced by increased anaerobic metabolism of muscle cells. The increase in lactic acid and drop in blood pH following violent exercise can cause death by asphyxiation in some fish (Black, 1958).

In the natural environment, many fish species hide after swimming stress until physiological parameters have reequilibrated. Other species live in fast water habitats that do not provide such an opportunity (Powers, 1972). Some, but by no means all, of these fast water species, have some hemoglobins with normal alkaline Bohr effects and one or more hemoglobin that are relatively unaffected by changes in pH. The sympatric Catostomid fishes mentioned earlier were the first example that illustrated this point. Catostomids are distributed so that only one species of a subgenus inhabits a given geographical region, though each species of one subgenus is usually found living with a member of the other subgenus (i.e., sympatric). Though distinct in number of morphological characteristics, the sympatric species, C. (Pantosteus) clarkii and C. (Catostomus) insignis, can also be distinguished from one another by the presence or absence, respectively, of cathodal components in the electrophoretic pattern of their hemoglobins (Powers, 1972). The anodal hemoglobins from both species show similar Hb-O_2 affinities and similar Bohr and Root effects. However, the cathodal hemoglobins isolated from C. clarkii do not have a significant sensitivity to changes in pH (Powers, 1972).

The NH protons in the imidazolium ring of COOH-terminal histidine in the β chains and NH_2-termini of the α chains are known to be largely responsible for the Bohr effect in mammalian hemoglobins (Kilmartin and Rossi-Bernardi, 1969; Kilmartin and Wooton, 1970; Perutz et al., 1969). As pointed out in 1972, the anodal catostomid hemoglobins had Tyr-His at the COOH-termini of the β chains, while the cathodal (non-Bohr effect) hemoglobin had Tyr-Phe. Moreover, the α chains had blocked NH_2-termini. Thus, the reduced pH sensitivity was at least partly explained by these amino acid substitutions (Powers, 1972; Powers and Edmundson, 1972b). We predicted (Powers, 1972) that similar adaptations should exist in other ecologically equivalent species and it was reassuring when Barra et al. (1973) found similar amino acid substitutions in the trout I hemoglobin.

I postulated (Powers, 1972) that the cathodal hemoglobins, without the Bohr effect, provided an emergency back-up system to allow continued swimming in the fast water habitat during transient acidosis following emergency exertion (e.g., to escape a predator). Since I correctly predicted similar adaptation for other hyperactive fish, this hypothesis (Powers, 1972) has become widely adopted as a "well-known phenomenon of hyperactive fish". Actually, this hypothesis concerns transient acidosis that is applicable to a variety of aquatic ecotypes including some but not necessarily all hyperactive fish. This point and its precedent is often ambiguously and/or incorrectly referenced.

2. The Organic Phosphate Effect

The major organic phosphate in fish erythrocytes is either adenosine triphosphate (ATP) (e.g., Gillen and Riggs, 1971) or guanosine triphosphate (GTP) (Geohegan and Poluhowich, 1974), while in mammalian red blood cells it is 2,3-diphosphoglycerate (2,3 DPG). The effects of ATP and GTP on the oxygen binding properties of fish hemoglobins are similar to those of 2,3 DPG on mammalian hemoglobins. Furthermore, the influence of ATP on both the Bohr effect and on the pH dependence of the subunit cooperativity indicates that the reactions of many fish hemoglobins with oxygen, organic phosphates, and protons are linked functions (see earlier discussion). In addition to its allosteric effect on oxygen affinity, organic phosphate has an important influence on the Donnan distribution of protons across the erythrocyte membrane (Wood and Johansen, 1972). Several workers have demonstrated that adaptive changes in fish hemoglobin-oxygen affinity during acclimation to low oxygen or increased temperature are inversely related to changes in red blood cell organic phosphate concentrations (e.g., Greaney and Powers, 1978; Powers, 1974; Wood and Johansen, 1972; Wood et al., 1975).

A number of studies have shown that one mole of organic phosphate binds for each mole hemoglobin tetramer and that the NH_2-termini of the β chains are involved in ionic interactions with the organic phosphate. X-ray diffraction studies have confirmed that one molecule of organic phosphate (i.e., 2,3 DPG) binds to human deoxyhemoglobin on the diad axis of the tetramer (Arnone, 1972). The stereochemistry of the bound 2,3 DPG appears to complement certain basic residues in the central cavity of the deoxytetramer. Presumably, salt bridges are formed with the NH_2-termini of the β chains (Val, β 1), and the imidazole side chains (His: β 2, β 143) of both β chains, as well as the ϵ-aminogroup of Lys, β 82, from one of the β chains. The homologous residues for some fish hemoglobins are: Val (β 1), Glu (β 2), Lys (β 82), and Arg (β 143) (Bossa et al., 1978; Grujic-Injac et al., 1980; Powers and Edmundson, 1972b). Assuming that the same amino acid residues are involved in the binding of organic phosphate by fish hemoglobins, the only residues above that could be titrated over a physiological pH range are the two valyl residues of the β chain NH_2-termini and the β 82 lysyl residues.

Since ATP is the major organic phosphate modifier of many fish hemoglobins, the binding of ATP to representative fish hemoglobins has been investigated (Greaney et al., 1980). Since there is a strong linkage between proton and ATP binding, an analysis of the effect of pH on ATP affinity has been done for Carp HbI deoxyhemoglobin (see Fig. 2A), (Greaney et al., 1979). The study indicates that Carp deoxyhemoglobin I binds three protons when ATP^{4-} is bound, but only two protons when ATP^{3-} is bound. Presumably, two of the three are the α-amino groups of the β 1 valyl residues. The third could be one of the β 82 lysyl- ϵ-amino groups or perhaps another residue whose involvement is yet to be elucidated.

Since the binding of ATP^{4-} involves a greater ionic interaction than that of ATP^{3-}, one would expect significant differences in the respective hemoglobin-ATP affinities. We have shown that hemoglobin's affinity for the fully ionized species (ATP^{4-}) is approximately six orders of magnitude greater than that for the partially ionized form (ATP^{3-}). Therefore, it is important to consider the various ionic species of organic phosphates when oxygen equilibria studies are designed. However, as has been the case for the ligand-linked effect of subunit dissociation, these and other important parameters are usually ignored.

IV. ADAPTATIONS TO TEMPERATURE CHANGES

As temperature increases, the availability of oxygen in water decreases because: (1) the solubility of oxygen decreases with increasing temperature (Henry's law) and (2) elevated biological activity (e.g., bacterial and planktonic) at high temperatures often reduce oxygen to below saturation. Consequently, at higher temperatures fish require more oxygen but less is available.

Fish respond to this dilemma by a combination of strategies. For example, ventilation volume and heart rate increase directly with temperature (reviewed by: Prosser, 1973). Fish increase the oxygen carrying capacity of the blood by increasing hematocrit (Cameron, 1970; DeWilde and Houston, 1967; Greaney and Powers, 1977; Houston and Cyr, 1974; Powers, 1974; Powers and Powers, 1975). However, this phenomenon may be primarily a seasonally induced response to oxygen rather than a response to temperature per se (Denton and Yousef, 1975; Greaney and Powers, 1977). In some species there are changes in intracellular levels of allosteric modifiers (e.g., Greaney and Powers, 1977) which, in turn increase hemoglobin oxygen affinity at higher temperatures.

The effect of temperature on the oxygen equilibria of fish hemoglobin has been reviewed (Johansen and Lenfant, 1972; Johansen and Weber, 1976; Riggs, 1970). Although the number of species examined is limited, the functional properties of teleost hemoglobins can be divided into three major categories as suggested by Weber et al. (1976). Class I contains species with one or more hemoglobins all of which are sensitive to both temperature and pH (e.g., Bonaventura et al., 1974; Gillen and Riggs, 1971; Gillen and Riggs, 1972; Mied and Powers, 1977; Weber, 1975). Class II has species with multiple components, some of which are functionally similar to the Class I hemoglobins, while other components are not strongly affected by temperature and pH (e.g., Binotti et al., 1971; Hashimoto et al., 1963; Powers and Edmundson, 1972a; Powers, 1972; Powers, 1974; Wyman et al., 1977). Class III fishes have hemoglobins that are pH sensitive , but temperature insensitive (e.g., Anderson et al., 1973; Rossi-Fanelli and Antonini, 1960).

These studies have shown that among fish hemoglobins the apparent enthalpy (Δ H) of oxygenation varies from zero to about 16 kcal per mole.

Barcroft and King (1901) were the first to show that an increase in temperature will decrease the affinity of blood for oxygen. Temperature affects a wide variety of biochemical phenomena that can either directly or indirectly influence oxygen equilibria. For example, temperature affects erythrocyte membrane fluidity, blood pH, the binding constants of organic phosphates and the activity coefficients of a number of intra- and extra-cellular components. The overall enthalpy, ΔH, of blood oxygen equilibria represents both the intrinsic heat of oxygen binding to hemoglobin and contributions due to other ligand-linked processes. One such process is associated with proton equilibria of amino acid side chains (i.e., the Root and Bohr effects). At alkaline pH (i.e., pH 9-10) the Root and Bohr effects are essentially inoperative. Thus, data collected on stripped hemoglobins at alkaline pH, over a range of temperatures, provide information primarily on the intrinsic thermodynamic parameters of the hemoglobins per se. Under such conditions, enthalpies are very similar for a number of hemoglobins of fish from a variety of thermal environments, particularly the Class I hemoglobins and Class II thermal sensitive multi-hemoglobins components (Powers et al., 1979b).

The strongest evidence for unique reduced thermal sensitivities of Hb-O_2 affinity has been provided by several species of tuna and some of their predators. For example, the blue fin tuna (Thunnus thynnus) swims through a variety of thermal environments and has hemoglobins that show little or no temperature sensitivities (Rossi-Fanelli and Antonini, 1968). This hemoglobin adaptation could be an adaptation to the animals internal heterothermy (Carey et al., 1966, 1969, 1971) and fluctuating thermal environment.

There are many poikilothermic fish that live in annually fluctuating thermal environments and have hemoglobins with reduced temperature sensitivity (e.g., Brunori, 1975; Powers, 1980). In every case, the reduction in temperature dependence results from the presence of one or more unique hemoglobin components which are relatively insensitive to temperature. These hemoglobins are usually accompanied by other components that are significantly affected by temperature.

Since Amazonian fish live in areas where the water temperature varies only a few degrees over the period of a year, our finding that their hemoglobins demonstrate a significant temperature effect is consistent with Johansen ans Lenfant's (1972) reasoning for blood oxygen affinity of animals in "constant thermal environments". However, most temperate fish species have hemoglobins with thermal sensitivity similar to those of their tropical counterparts (see Powers et al., 1979b; Powers, 1980, and literature there in) which is not consistent with Johansen and Lenfant's view. Perhaps the best example is the euryhaline minnow Fundulus heteroclitus. Fundulus have a limited home range of 36 meters (Lotrich, 1975) and live in one of the steepest annual thermal gradients in the world (Powers and Place, 1978). Moreover, their intertidal habitat is characterized by significant daily temperature changes ($\Delta T = 10$ °C per day and as much as 40 °C per year). Since Fundulus hemoglobins are just as thermally sensitive as those of fish from "constant thermal environments" a generalized theory regarding hemoglobin, habitat, and thermal stability, does not appear to be justified. Thus while a selective advantage for a thermally insensitive hemoglobin could conceivably be made for some unique species, the generalized evolutionary development of reduced thermal sensitivity of hemoglobin for fish in fluctuating thermal environments is not compelling. If evolution has favored a decrease in temperature sensitivity of $Hb-O_2$ affinity in fish experiencing large fluctuations, then it must be primarily associated with the regulation of intracellular pH, the levels and types of organic phosphates, and other ligand-linked phenomena rather than selection of hemoglobins per se. Thus, the thermal adaptation must be primarily at the erythrocyte level rather than the intrinsic enthalpy of hemoglobin oxygen binding.

Greaney and Powers (1977) demonstrated that F. heteroclitus adapted to thermal changes by changing ATP concentrations within erythrocytes. However, studies on other fish species (Powers, unpublished) indicate that the strategy employed by Fundulus is not applicable to all fishes. At least two fresh water fish and one salt water species show no significant thermal compensation.

Since a physical dependence exists between oxygen solubility and water temperature (oxygen concentration is 12.5 p.p.m. at 10 °C compared with 7.5 p.p.m. at 30 °C), our observation that F. heteroclitus lowers red blood cell ATP when acclimatized to elevated temperatures (Greaney and Powers, 1978) prompted us to ask if this response was triggered by reduced oxygen (due to the increased temperature), increased temperature, or both of these variables. Therefore, fish were acclimated to various temperatures (10 °C, 22 °C, and 30 °C), but with oxygen concentration maintained constant (about 7 p.p.m.) at each temperature. These animals also decreased their erythrocyte ATP with increased temperature. Moreover, fish acclimated to 10 °C, but in air saturated water (12.5 p.p.m.) showed the same red blood cell ATP levels as those maintained at 10 °C but with 7 p.p.m. oxygen. These data demonstrate that the ATP responses was elicited by increased temperature alone and was independent of dissolved oxygen levels in the range 7-12.5 p.p.m.

V. STRATEGIES OF ADAPTATING TO HYPOXIA

There are numerous strategies by which fish are able to maintain respiratory homeostasis when environmental oxygen is reduced. Perhaps the most common is to seek out a more favorable environment. Species who remain in oxygen poor environments have: (1) immediate, (2) intermediate, and (3) long-term responses.

A. Immediate Response to Hypoxia

The immediate response to an oxygen poor environment is an increase in heart rate and ventilation volume (Prosser, 1973). In addition, fish will often "gulp" air and/or primarily utilize the water at the air-interface which has the greatest oxygen content. In addition, fish are able to rapidly increase or decrease blood oxygen affinity by employing hormonal and/or other factors in the serum (Powers et al., 1985). For example, Dalessio et al. (1984) have examined the effect of various concentrations of epinephrine, propranolol and phenoxybenzamine on the oxygen affinity of F. heteroclitus blood. Epinephrine decreased oxygen affinity of whole blood both in vivo and in vitro while it increased oxygen affinity of saline washed cells. Its effects were exerted via the β-adrenergic receptor and were not dose dependent in vivo. The oxygen affinity of whole blood was substantially higher than that of saline washed cells. Addition of fresh serum to washed cells increased their oxygen affinity to values identical to those found for whole blood. These findings indicate that epinephrine and an unidentified component in fish serum substantially effect Hb-O_2 affinity of blood.

B. Intermediate Response to Hypoxia

Intermediate responses are generally activated after several hours of hypoxic conditions and last many days or until longterm compensation is achieved. These responses include increasing hematocrit by retaining serum in muscle tissues (Cameron, 1970), or releasing stored erythrocytes from the spleen reducing intraerythrocyte organophosphate concentrations (e.g., Greaney and Powers, 1977, 1978; Wood and Johansen, 1972), changing pH, and changing the ionic micro-environments of erythrocytes (Houston and Mearow, 1979). Such attempts to maintain oxygen delivery to respiring tissues is usually accompanied by large fluctuations in various enzyme activities during metabolic readjustment (Greaney et al., 1980).

During the intermediate responses described above, fish synthesis new erythrocytes so that the total oxygen carrying capacity of the blood is increased. Eventually, a new steady-state between new cells, enzyme levels, organophosphates and hemoglobin function is achieved. This level and the balance is different for each fish species. I shall now describe strategies employed by some representative species, but there does not appear to be a common adaptive strategy. The long-term evolutionary responses will not be addressed in this paper.

C. Whole Animal Level Responses

When exposed to low oxygen environments for several days, both mammals and fish (Krogh and Leitch, 1919; Johansen, 1970) increase the oxygen carrying capacity of their blood. This is accomplished by a number of strategies. The common mechanisms between mammals and fish are: increased hematocrit, increased hemoglobin content, and increased blood buffering capacity. On the other hand, mammals and fish differ considerably in other aspects of their response to low oxygen. Mammals decrease the affinity of their hemoglobin for oxygen. Hypoxic fish, in contrast, increase hemoglobin-oxygen affinity as characterized by a decrease in the P_{50} of their corresponding oxygen-saturation curves (Wood and Johansen, 1972). For example, mammals acclimated to high altitudes, increase levels of 2,3 DPG in their erythrocytes (Lenfant et al., 1968). Since mammals typically live in an oxygen-rich environment, it has been suggested that this response might be best suited to the more commonly encountered forms of low altitude hypoxia, such as chronic hypoxemia (Eaton, 1974; Gerlach and Duhm, 1972). The response of water-breathing vertebrates to chronic hypoxia is quite different (Powers, 1974). For example, eels have been shown to decrease erythrocyte ATP and increase $Hb-O_2$ affinity when acclimated to low environmental oxygen (Wood and Johansen, 1972). In additon, F. heteroclitus acclimated to hypoxic conditions lower red cell ATP by as much as 40% and increase the percent hematocrit which presumably increases the oxygen carrying capacity of the blood (Fig. 3) (Greaney and Powers, 1978). Moreover, there were concomitant increases in serum lactate and a decrease in blood pH.

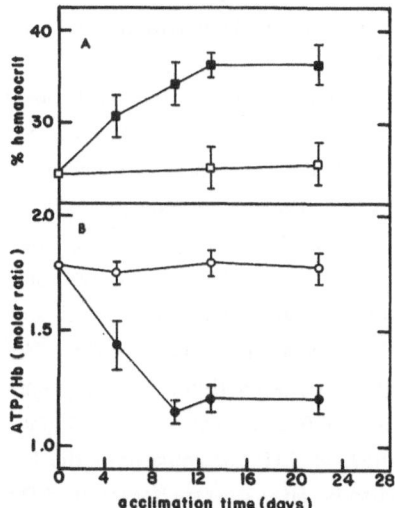

FIG. 3. Fundulus heteroclitus hypoxia studies: (A) Time course of % blood hematocrit of hypoxic (■) and normoxic (□) fish. (B) Time course of acclimating of F. heteroclitus to hypoxic (●) and normoxic (○) conditions at 22 °C. Dissolved oxygen values were 0.2 - 2.0 parts per million (ppm) for hypoxic and 8.5 - 9.0 ppm for control fish. Red cell ATP for this experiment and all others described in this paper were determined using the firefly luciferase assay. All points represent averages of 6-7 fish. Bars indicate ± standard error of the mean (S.E.M.).

Cellular level - We predicted that if the control mechanism was directed at the erythrocyte level, then fish red blood cells should decrease ATP, in vitro, under anoxic conditions. Consistent with our in vivo observation, we found that anaerobic F. heteroclitus red cells significantly lowered their ATP levels in vitro. Since the nucleated erythrocytes of fish possess mitochondria, we reasoned that this response may be mediated by way of a decrease in oxidative phosphorylation. This hypothesis was supported when aerated cells were incubated in the presence of low concentrations of cyanide. This inhibitor of aerobic respiration reduced intracellular ATP levels similar to those found in the anoxic cells (Greaney and Powers, 1977, 1978). These data have been confirmed by red cell oxygen consumption studies (Powers, 1983). Thus it seems that one reason fish have retained a functional oxygen-consuming electron transport system in the mitochondria of their red blood cells is to control the levels of hemoglobin allosteric effector. Our current studies on the intermediate metabolism of Fundulus erythrocytes should provide this mechanism.

VI. PHYSIOLOGICAL CORRELATION BETWEEN LACTATE DEHYDROGENASE GENOTYPE AND HEMOGLOBIN FUNCTION IN KILLIFISH

During our hypoxic studies, we observed significant variability in red cell ATP concentrations among individual fish. We wondered whether this variability could be the result of a genetic component. Consequently, we analyzed F. heteroclitus for red cell ATP, concomitantly screening for a series of enzyme variants. Not only were ATP levels under genetic control, but they were also highly correlated with lactate dehydrogenase (LDH) phenotype (Powers et al., 1979c).

There are three major electrophoretically distinguishable phenotypes of the heart-type LDH (the LDH-B locus) in F. heteroclitus (Place and Powers, 1978). The polymorphism is due to two co-dominant alleles which exhibit a dramatic north-south cline in gene frequency along the Atlantic coast of the U.S. (Powers and Powers, 1975; Powers and Place, 1978). The phenotypes are LDH-$B^a B^a$, LDH-$B^a B^b$, and LDH-$B^b B^b$. The B locus is the only LDH expressed in the red blood cells of this fish. When intraerythrocyte ATP levels are compared for individuals of different LDH-B phenotypes, there was significant association. As the intraerythrocyte ATP/Hb ratio is different for each of the LDH-B phenotypes, we expected to find differences in hemoglobin-oxygen affinity. In agreement with this expectation, LDH-$B^a B^a$ homozygotes with the lowest ATP/Hb ratio had the highest oxygen affinity and LDH-$B^b B^b$ homozygotes with the highest ATP/Hb value have the lowest oxygen affinity (Powers et al., 1979c). This was found to be an important factor in differential developmental rates and swimming abilities (DiMichele and Powers, 1982a; 1982b; 1984) as briefly described below.

A. Development Rate and Hatching Time

DiMichele and Taylor (1980; 1981) have shown that respiratory stress triggers the hatching mechanism in F. heteroclitus. In view of this, we (DiMichele and Powers, 1982a) reasoned that the hatching rates of LDH-B phenotypes should differ because of differences in hemoglobin-oxygen affinity. Consistent with that expectation we found that F. heteroclitus embryos hatched at rates that were highly correlated with LDH-B phenotype. LDH-B^{a}B^a individuals hatched before LDH-B^{b}B^b phenotypes and the heterozygotes (LDH-B^{a}B^b) had an intermediate hatching distribution.

We found (DiMichele and Powers, 1982a) that LDH B^{a}B^a eggs dominated the first three days of the hatching period and the LDH-B^{b}B^b eggs dominated the last half. The LDH-B^{a}B^b eggs essentially hatched over the entire time span. The overall mean hatching times for offspring of 20 random crosses were: 11.9 days for the LDH-B^{a}B^a phenotype, 12.4 days for the LDH-B^{a}B^a phenotype, and 12.8 days for the LDH-B^{b}B^b phenotype.

The time of hatching is very important in Fundulus populations because of its unique reproductive strategy. The eggs of F. heteroclitus are laid in empty mussel shells or between the leaves of the marsh grass, Spartina alterniflora (Taylor et al., 1977). Under these conditions, the eggs incubate in air for most of their developmental period. Hatching occurs when eggs laid on one spring tide are immersed in water by the following spring tide. As water covers the eggs, there is a drop in environmental oxygen at the egg surface which is hatching cue for the embryo (DiMichele and Taylor, 1980). Hatching at the correct time would seem to be important for survival of the fry. Therefore, overall plasticity in hatching times may be important in protecting F. heteroclitus populations that live under variable environmental conditions. Our data suggest that premature hatching cues (e.g., rainstorms) would select mostly against LDH-B^{a}B^a individuals, while late hatching (i.e., after the tide has retreated) would primarily select against the LDH-B^{b}B^b phenotypes. This argument is particularly compelling in the light of the recent finding of Meredith and Lotrich (1979) that the mortality of F. heteroclitus in age class zero (eggs to fry of 59 mm) is greater than 99.5%.

If extremes in hatching time are selected against, then there would be a net heterozygote advantage in a variable or uncertain environment. Such an advantage would result in the maintenance of genetic variability at the LDH-B locus as well as a stability in gene frequency at those localities where such selection operates. This is consistent with the temporal stability in LDH-B gene frequencies at several localities (Powers and Place, 1978). However, spatial changes in the gene frequencies for a number of loci are probably due to other physical, biological and stochastic factors.

Since erythrocyte ATP concentrations are correlated with LDH-B genotype and ATP regulates hemoglobin-oxygen affinity, the simplest interpretation is that hypoxia induced hatching of LDH-B variants results from functional difference between LDH-B allozymes which affect ATP levels. However, DiMichele and Powers (1984a, 1984b) have shown that developmental rates of the various genotypes are also different.

B. Physiological Basis for Swimming Endurance Differences LDH-B Genotypes

Our analysis of purified LDH-B allelic isozymes (Place and Powers, 1979, 1984a, 1984b) indicated that the greatest catalytic differences between LDH-B^{a}B^a and LDH-B^{b}B^b existed at low temperature (10 °C) while no significant difference exists at 25 °C. Thus, if the LDH-B enzyme has a direct influence on erythrocyte ATP concentration, differences in ATP and blood oxygen affinity should only exist at acclimation temperatures below 25 °C. Furthermore, since organic phosphate amplifies the Bohr effect of F. heteroclitus hemoglobins (Mied and Powers, 1977), these phenomena should be exaggerated at low blood pH values, like those produced during swimming performance experiments. DiMichele and Powers (1982b) have reported that swimming performance is highly correlated with genetic variation at the LDH-B locus for Fundulus acclimated to 10 °C while no such differences exist for 25 °C acclimated fish.

After an acclimation period, fish of each of the two homozygous LDH-B phenotypes were swum to exhaustion in a closed water tunnel. The exhausted fish were sacrificed immediately and appropriate biochemical and physiological parameters determined (Table 2).

Among resting fish acclimated to 10 °C, hematocrit, blood pH, blood oxygen affinity, serum lactate, liver lactate, and muscle lactate were not significantly different between the two LDH-B homozygous phenotypes (Table 2). Exercising fish, acclimated to 10 °C, to the point of fatigue caused a significant change in all of these parameters. The data in Table 2 show that the LDH-B B^{bb} phenotype was able to sustain a swimming speed 20% higher than that of LDH-B^{a}B^a phenotype. Blood oxygen affinity, serum lactate and muscle lactate also differed between the two phenotypes. Since the rate of lactate accumulation was the same for the LDH-B phenotypes, LDH-B^{b}B^b fish accumulated more lactate in the blood and muscle simply because they swam longer.

In an analysis of the binding of ATP to Carp deoxyhemoglobin, Greaney et al. (1979) have shown that the organophosphate-hemoglobin affinity constants change by two orders of magnitude between pH 8 and pH 7. The same general phenomenon appears to be true for F. heteroclitus hemoglobins (Greaney and Powers, unpublished; Powers, 1980). In resting Fundulus at 10 °C, the blood pH was about 7.9 (Table 2). At this pH, the difference in erythrocyte ATP between LDH-B phenotypes (ATP/Hb were 1.65 $\pm$ 0.12 and 2.11 $\pm$ 0.22 for LDH-B^{a}B^a and LDH-B^{b}B^b respectively) is not reflected as a significant difference in blood oxygen affinity. However, as blood pH falls with increasing exercise, the organophosphate-hemoglobin affinity constant increases, and differences in oxygen affinity between homozygous LDH-B genotypes becomes apparent (Fig. 4). As blood pH is lowered, ATP amplifies the dissociation of oxygen from F. heteroclitus hemoglobin; the more ATP, the greater the effect. This difference is translated into a differential ability to deliver oxygen to muscle tissue which in turn affects swimming performance (DiMichele and Powers, 1982b).

This phenomenon is particularly important because it illustrates how an enzyme that is not involved in respiration can indirectly effect the availability of oxygen and thus the ability to perform work.

TABLE 2. Biochemical and physiological parameters determined on two homozygous LDH-B phenotypes of Fundulus heteroclitus at rest and after swimming to exhaustion in a water tunnel (from DiMichele and Powers, 1982b)

	10 °C			25 °C		
	LDH-B^{a}B^a	LDH-B^{b}B^b	P	LDH-B^{a}B^a	LDH-B^{b}B^b	P
Resting fish						
Hematocrit	23 ± 1	24 ± 1	N.S.	24 ± 1	25 ± 2	N.S.
Blood pH	7.87 ± 0.05	7.84 ± 0.04	N.S.	7.40 ± 0.05	7.48 ± 0.04	N.S.
P$_{50}$ (mmHg)	4.2 ± 0.2	3.8 ± 0.2	N.S.	5.0 ± 0.3	4.7 ± 0.2	N.S.
Serum lactate (mM)	1.82 ± 0.28	1.37 ± 0.15	N.S.	2.6 ± 0.69	4.5 ± 0.86	N.S.
Liver lactate (μmole/g)	0.390 ± 0.055	0.383 ± 0.44	N.S.	1.8 ± 0.42	1.53 ± 0.39	N.S.
Muscle lactate (μmole/g)	7.93 ± 0.75	8.17 ± 1.02	N.S.	11.8 ± 2.2	12.5 ± 1.6	N.S.
Exercised fish						
Critical swim speed (body lengths per second)	3.6 ± 0.12	4.3 ± 0.1	*	5.6 ± 0.3	5.8 ± 0.3	N.S.
Hematocrit	30 ± 2	35 ± 2	N.S.	36 ± 1	38 ± 2	N.S.
Blood pH	7.24 ± 0.04	7.15 ± 0.05	N.S.	7.12 ± 0.03	7.09 ± 0.09	N.S.
P$_{50}$ (mmg)	6.57 ± 0.5	9.1 ± 0.5	*	7.4 ± 0.6	7.2 ± 0.5	N.S.
Serum lactate (mM)	12.19 ± 1.21	16.29 ± 0.79	*	23.4 ± 2.5	17.5 ± 1.6	N.S.
Liver lactate (μmole/g)	1.39 ± 0.12	1.56 ± 0.17	N.S.	6.6 ± 1.2	5.0 ± 0.6	N.S.
Muscle lactate (μmole/g)	17.08 ± 1.86	24.01 ± 1.46	*	23.2 ± 4.8	20.7 ± 1.6	N.S.

* Significant at $P < 0.05$

118

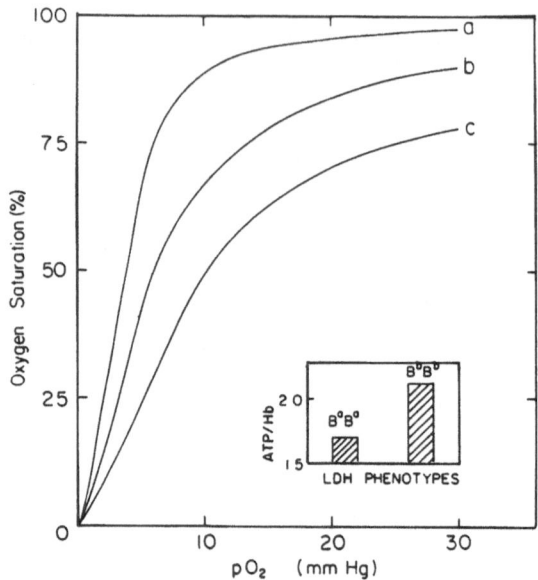

FIG. 4 Oxygen equilibrium curves for whole blood of <u>Fundulus</u> <u>heteroclitus</u> acclimated to 10 °C, as determined with an oxygen dissociation analyzer (Aminco). The ordinate is the percentage saturation of hemoglobin by oxygen, and the abscissa is the partial pressure of oxygen (PO_2). Oxygen equilibrium curves of blood from (a) resting fish of both LDH-B phenotypes, (b) LDH-B^{a}B^a swum to exhaustion, and (c) LDH-B^{b}B^b swum to exhaustion. The intraerythrocyte ratio of ATP to hemoglobin (ATP/Hb) in resting fish is 1.65 $\pm$ 0.12 for LDH-B^{a}B^a and 2.11 $\pm$ 0.22 for LDH-B^{b}B^b.

REFERENCES

Ackers GK (1979) Linked functions in allosteric proteins: an exact theory for the effect of organic phosphates on oxygen affinity of hemoglobin. Biochemistry 15: 3372-3380

Anderson ME, Olson JS, Gibson QH (1973) Studies on ligand binding to hemoglobins from teleosts and elasmobranches. J. Biol. Chem. 248: 331-341

Arnone A (1972) X-ray diffraction study of binding of 2,3 diphosphate glycerate to human deoxyhemoglobin. Nature 237:146-149

Barcroft J, King WOR (1909) The effect of temperature on the dissociation curve of blood. J. Physiol. Lond. 39: 374-384

Barra D, Bossa F, Bonaventura J, Brunori M (1973) Hemoglobin components from trout (<u>Salmo</u> <u>irideus</u>): Determination of the carboxyl and amino terminal sequences and their functional implications. FEBS Lett. 35: 151-154

Barr D, Petruzzelli R, Bossa F, Brunori M (1983) Primary structure of hemoglobin from trout (<u>Salmo</u> <u>irideus</u>) amino acid sequence of the β-chain of trout HbI. Biochim. Biophys. Acta 742: 72-77

Benesch R, Benesch RE (1967) The effect of organic phosphates from the human erythrocyte on the allosteric properties of hemoglobin. Biochem. Biophys. Res. Commun. 26:162-167

Benesch R, Benesch RE, Yu CI (1968) Reciprocal binding of oxygen and diphosphoglycerate by human hemoglobin. Proc. Natl. Acad. Sci. 59: 526-532

Benesch RE, Benesch R (1970) The reaction between diphosphoglycerate and hemoglobin. Fed. Proc. 29: 1101-1104

Binotti S, Giovenco B, Giardina B, Antonini E, Brunori M, Wyman J (1971) Studies of the functional properties of fish hemoglobins. II. The oxygen equilibrium of the isolated hemoglobin components from trout blood. Arch. Biochem. Biophys. 142: 274-281

Black EC (1958) Hyperactivity as a letahl factor in fish. J. Fish Res. Bd Can. 15: 573-586

Bonaventura J, Bonaventura C, Sullivan B (1975) Hemoglobins and hemocyanins: Comparative aspects of structure and function. J. Exp. Zool. 194: 155-174

Bonaventura J, Gillen R, Riggs A (1974) The hemoglobin of the Crossopterygerian fish Latimeria chalumnae (Smith) - Subunit structure and oxygen equilibrium. Arch. Biochem. Biophys. 163: 728-734

Bossa F, Barra D, Coletta M, Martini F, Liverzani A, Petruzzelli R, Bonaventura J (1976) Primary structure of haemoglobins from trout (Salmo irideus). Partial determination of amino acid sequence of Hb trout IV. FEBS Lett. 64: 76-80

Bossa F, Barra D, Petruzzelli R, Martini F, Brunori M (1978) Primary structure of hemoglobin from trout (Salmo irideus). Biochim. Biophys. Acta 536: 298-305

Braunitzer G, Rodewald K (1980) Die Sequenz der α- and β-Ketten des Hamoglobins des Goldfisches (Carassius auratus). Hoppe-Seyler's Z Physiol. Chem. 361: 587-590

Briehl RW (1963) A relation between O_2-Hb equilibria and aggregation of subunits in lamprey hemoglobins. J. Biol. Chem. 238: 2361-2366

Brunori M (1966) Bohr effect in hemoglobin from Thunnus thynnus. Arch. Biochem. Biophys. 144: 195-210

Brunori M, Bonaventura J, Bonaventura C, Giardina B, Bossa F, Antonini E (1973) Hemoglobins from trout: structural and functional properties. Mol. Cell. Biochem. 1: 189-196

Brunori M (1975) Molecular adaptation to physiological requirements: The hemoglobin system of trout. Curr. Topics Cell. Regul. 9:1-39

Cameron JN (1970) The influence of environmental variables on the haematology of the pinfish (Lagodon rhomboides) and striped mullet (Mugil cephalus). Comp. Biochem. Physiol. 32: 175-192

Carey FG (1973) Fishes and Warm Bodies. Sci. Amer. 228: 36-44

Carey FG, Teal JM (1966) Heat conservation in tuna muscle. Proc. Natl. Acad. Sci. USA 56: 1464-1469

Carey FC, Teal JM (1969) Regulation of body temperature by the bluefin tuna. Comp. Biochem. Physiol. 28: 205-213

Carey FG, Teal JM, Kanwisher JW, Lawson KD (1971) Temperature regulation in Tuna. Amer. Zool. 11: 137-145

Chanutin A, Curnish RR (1964) Factors influencing the electrophoretic patterns of red cell hemolysates analyzed in cacodylate buffers. Arch. Biochem. Biophys. 106: 433-439

Chanutin A, Curnish RR (1967) Effect of organic and inorganic phosphates on the oxygen equilibrium of human erythrocytes. Arch. Biochem. Biophys. 121: 96-102

Chien JC, Mayo KH (1980) Carp hemoglobin. J. Biol. Chem. 225: 9790-9799

Dalessio P, DiMichele L, Powers DA (1984) Extracellular control of erythrocyte oxygen affinity in teleosts. Amer. Zool. 24: 120A

Denton JE, Yousef MF (1975) Seasonal changes in hematology of rainbow trout, Salmo gairdneri. Comp. Biochem. Physiol. 51A: 151-153

DeWilde MA, Houston AH (1967) Haematological aspects of thermoaccclimatory process in the rainbow trout Salmo gairdneri. J. Fish Res. Bd Can. 24: 2267-2281

DiMichele L, Powers DA (1982a) LDH-B genotype specific hatching times of Fundulus heteroclitus embryos. Nature 260: 563-564

DiMichele L, Powers DA (1982b) The physiological basis for swimming endurance differences between LDH-B genotypes of Fundulus heteroclitus. Science 216: 1014-1016

DiMichele L, Powers DA (1984a) The relationship between oxygen consumption rate and hatching in Fundulus heteroclitus. Physiol. Zool. 57: 46-51

DiMichele L, Powers DA (1984b) Developmental and oxygen consumption differences between LDH-B genotypes of Fundulus heteroclitus and their effect on hatching time. Physiol. Zool. 57: 52-56

DiMichele L, Taylor MH (1980) The environmental control of hatching in Fundulus heteroclitus. J. Exp. Zool. 214: 181-187

DiMichele L, Taylor MH (1981) The mechanism of hatching in Fundulus heteroclitus: Development and physiology. J. Exp. Zool. 217: 73-79

Eaton JW (1974) Oxygen affinity and environmental adaptation. Ann. N.Y. Acad. Sci. 241: 491-497

Fermi G, Perutz MF (1981) Haemoglobin and myoglobin. Atlas of biological structures, vol. II, Clarendon, Oxford

Fyhn U, Fyhn H, Davis J, Powers DA, Fink WT, Garlick R (1979) Hemoglobin heterogeneity in Amazonian fishes. Comp. Biochem. Physiol. 62A: 39-66

Fyhn U, Sullivan B (1974) Hemoglobin polymorphism in fishes. I. Complex phenotypic patterns in the toadfish. Biochem. Genetics 11: 373-385

Fyhn U, Sullivan B (1975) Elasmobranch hemoglobins: dimerization and polymerization in various species. Comp. Biochem. Physiol. 50B: 119-129

Geoghegan WD, Poluhowich JJ (1974) The major erythrocyte organic phosphates of the American eel Anguilla rostrata. Comp. Biochem. Physiol. 49B: 281-290

Gerlach E, Duhm J (1972) 2,3 DPG Metabolism of red cells: Regulation and adaptative changes during hypoxia. In: Astrop P, Roth M (eds) Oxygen Affinity of Hemoglobin and Red Cell Acid Base Status. Academic Press, Munksgaard, Copenhagen, Denmark

Gillen RG, Riggs A (1971) The hemoglobins of a freshwater teleost Cichlasoma cyanoguttatum: The effects of phosphorylated organic compounds upon oxygen equilibria. Comp. Biochem. Physiol. 38B: 585-595

Gillen R, Riggs A (1972) Structure and function of the hemoglobins of the carp, Cyprinus carpio. J. Biol. Chem. 247: 6039-6046

Greaney GS, Powers DA (1977) Cellular regulation of an allosteric modifier of fish hemoglobin. Nature 270: 73-74

Greaney GS, Powers DA (1978) Allosteric modifiers of fish hemoglobins: In vitro and in vivo studies of the effect of ambient oxygen and pH on eryhtrocyte ATP concentrations. J. Exp. Zool. 203: 339-350

Greaney GS, Hobish MK, Powers DA (1980a) The effects of temperature and pH on the binding of ATP to Carp (Cyprinus carpio) deoxyhemoglobin (HbI). J. Biol. Chem. 255: 445-453

Greaney GS, Place AR, Cashon RE, Smith G, Power DA (1980b) Time course of changes in enzyme activities and blood respiratory properties of killifish during long-term acclimation to hypoxia. Physiol. Zool. 53: 136-144

Grujic-Injac B, Braunitzer G, Stangl A (1979) Die Aminosauresequenz der β-ketten der beiden Hauptkomponenten des Karpfen Haemoglobin. Hoppe-Seyler's Z Physiol. Chem. 360: 609-612

Grujic-Injac B, Braunitzer G, Stangl A (1980) Die Sequenz der $β_A$ and $β_B$ Ketten der Haemoglobine des Karpfens (Cyprinus carpio L.). Hoppe-Seyler's Z Physiol. Chem. 361: 1629-1639

Hashimoto K, Yamaguchi Y, Matsuura F (1960) Comparative studies on two hemoglobins of salmon IV. Oxygen dissociation curve. Bull. Japan Soc. Sci. Fish 26: 827

Hilse K, Braunitzer G (1968) Die Aminosauresequenz der α-ketten der beiden Hauptkomponenten des Karpfen Hemoglobin. Hoppe-Seyler's Z Physiol. Chem. 349: 433-450

Hjorth JP (1975) Molecular and genetic structure of multiple hemoglobins in the Eelpout, Zoarces vioiparus. Biochem. Genetics 13: 379-391

Hobish MK, Powers DA (1985) The effect of pH on the binding of DPG to human hemoglobin (Hb-A_o). J. Biol. Chem. (submitted)

Houston AH, Cyr D (1974) Thermoacclimatory variation in the hemoglobin system of goldfish (Carassius auratus) and rainbow trout (Salmo gairdneri). J. Exp. Biol. 61: 455-461

Houston AH, Mearow KM (1979) Temperature related changes in the ionic composition and (HCO_3^-) and (Na^+/K^+)-ATPase activities of the rainbow trout erythrocyte (submitted)

Iuchi I (1973) Ontogenetic expression of larval and adult hemoglobin phenotypes in the intergeneric salmoni hybrids. Comp. Biochem. Physiol. 44B: 1087-1101

Johansen K (1970) Air breathing in fishes. In: Hoar WS, Randall DJ (eds) Fish Physiology, Vol. IV, Academic Press, New York, pp. 361-411

Johansen K, Lenfant C (1972) A comparative approach to the adaptability of O_2-Hb affinity. In: Astrup P, Rorth M (eds) Oxygen Affinity of Hemoglobin and Red Cell Acid Base Status. Academic Press, Munksgaard, Copenhagen, Denmark

Johansen K, Weber RE (1976) On the adaptability of haemoglobin function to environmental conditions. In: Davies PS (ed) Perspectives in Experimental Biology. Pergamon Press, New York, pp. 212-234

Kilmartin JV, Rossi-Bernardi L (1969) Inhibition of CO_2 combination and reduction of the Bohr effect in hemoglobin chemically modified at its α-amino groups. Nature (London) 222: 1243-1246

Kilmartin JV, Wooten JF (1970) Inhibition of Bohr effect after removal of C-terminal histidines from haemoglobin β-chains. Nature 228: 766-767

Krogh A, Leitch I (1919) The respiratory function of blood in fishes. J. Physiol. Lond. 52: 288

Lenfant C, Johansen K (1968) Respiration in the African lungfish, Protopterus aethiopicus. I. Respiratory properties of blood and normal patterns of breathing and gas exchange. J. Exp. Biol. 49: 437-452

Lotrich VA (1975) Summer home range and movements of Fundulus heteroclitus (Pisces: Cyprinodontidae) in a tidal creek. Ecology 56: 191-198

Meredith WH, Lotrich VA (1979) Production dynamics of a tidal creek population of Fundulus heteroclitus (L.). Estuarine and Coastal Mar. Sci. 8: 88-118

Mied P, Powers DA (1978) Hemoglobins of the killifish Fundulus heteroclitus: Separation, characterization and a model for the subunit composition. J. Biol. Chem. 253: 3521-3528

Perutz MF, Lehmann H (1968) Molecular pathology of human hemoglobin. Nature 219: 902-909

Perutz MF, Muirhead H, Mazzarella L, Crowther RA, Greer J, Kilmartin JV (1969) Indentification of residues responsible for the alkaline Bohr effect in Haemoglobin. Nature 222: 1240-1243

Perutz MF (1984) Species adaptation in a protein molecule. Molecular Biology Evolution, Vol. I, n° 1, pp. 1-28

Place AR, Powers D (1978) Genetic basis for protein polymorphism in Fundulus heteroclitus. Biochem. Genetics 16: 577-591

Place AR, Powers DA (1979) Genetic variation and relative catalytic efficiencies: The LDH-B allozymes of Fundulus heteroclitus. Proc. Natl. Acad. Sci. USA 76: 2354-2358

Place AR, Powers DA (1984a) The lactate dehydrogenase (LDH-B) allozymes of Fundulus heteroclitus (Lin.): I. Purification and Characterization. J. Biol. Chem. 259: 1299-1308

Place AR, Powers DA (1984b) The lactate dehydrogenase (LDH-B) allozymes of Fundulus heteroclitus (Lin.): II. Kinetic analyses. J. Biol. Chem. 259: 1309-1318

Powers DA (1972) Hemoglobin adaptation for fast and slow water habitats in sympatric catostomid fishes. Science 177: 360-362

Powers DA (1974) Structure-function and molecular ecology of fish hemoglobins. Ann. N.Y. Acad. Sci. 241: 472-490

Powers DA, Powers DW (1975) Predicting gene frequencies in a natural population: A testable hypothesis. In: Markert C (ed) The Isozymes, IV Genetics and Evolution, Vol. IV, Academic Press, New York, pp. 63-84

Powers DA (1980) The molecular ecology of teleost fish hemoglobins: strategies for adapting to changing environments. Amer. Zool. 20: 139-162

Powers DA (1983) Adaptation of erythrocyte function during changes in environmental oxygen and temperature. In: Cossins AR, Sheterline P (eds) Cellular Acclimatisation to Environmental Change. Cambridge University Press, pp. 227-244

Powers DA, Edmundson AB (1972a) Multiple hemoglobins of catostomid fish. I. Isolation and characterization of the isohaemoglobin from Catostomus clarkii. J. Biol. Chem. 247: 6686-6693

Powers DA, Edmundson AB (1972b) Multiple hemoglobins of catostomid fish. II. The amino acid sequence of the major alpha chain from Catostomus clarkii hemoglobins. J. Biol. Chem. 247: 6694-6707

Powers DA, Place AR (1978) Biochemical genetics of Fundulus heteroclitus. I. Temporal and spatial variation in gene frequencies of Ldh-B, Mdh-A, Gpi-B and Pgm-A. Biochem. Genetics 16: 593-607

Powers DA, Fyhn HJ, Fyhn UFH, Martin JP, Garlick RL, Wood SC (1979a) A comparative study of the oxygen equilibria of blood from 40 genera of Amazonian fishes. Comp. Biochem. Physiol. 62A: 67-85

Powers DA, Martin JP, Garlick RL, Fyhn HJ (1979b) The effect of temperature on the oxygen equilibria of fish hemoglobins in relation to environmental thermal variability. Comp. Biochem. Physiol. 62A: 87-94

Powers DA, Greaney GS, Place AR (1979c) Physiological correlation between lactate dehydrogenase genotype and haemoglobin function in killifish. Nature 277: 240-241

Powers DA, Hobish MK, Greaney GS (1981) Rapid-rate equilibrium analysis of the interactions between organic phosphate and hemoglobins. In: Methods in Enzymology. Academic Press, Vol. 76, pp. 559-577.

Powers DA, DiMichele L, Place AR (1983) The use of enzyme kinetics to predict differences in cellular metabolism, developmental rate, and swimming performance between LDH-B genotypes of the fish Fundulus heteroclitus. In: Rattazzi MC, Scandalios JG, Whitt GS (eds) The Isozymes, Vol. X, Alan R Liss Inc. New York, pp. 147-170

Powers DA, Delassio P, DiMichele L (1985) The molecular ecology of Fundulus heteroclitus Hemoglobin-Oxygen affinity. Amer. Zool. (in press)

Prosser CL (1973) Comparative Animal Physiology, third edition. Saunders WB Co., Philadelphia PA

Reischl E (1976) The hemoglobins of the fresh water teleost Hoplias malabaria heterogeneity and polymorphism. Comp. Biochem. Physiol. 55B: 255-257

Riggs A (1970) Properties of fish hemoglobins. In: Hoar WS, Randall DJ (eds) Fish Physiology, Vol. IV. Academic Press, New York, pp. 209-252

Riggs A (1971) Mechanism of the enhancement of the Bohr effect in mammalian hemoglobins by diphosphoglycerate. Proc. Natl. Acad. Sci. 68: 2062-2065

Root RW (1931) The respiratory function of the blood of marine fishes. Biol. Bull. Mar. Biol. Lab. Woods Hole 61: 427-456

Rossi-Fanelli A, Antonini E (1961) Oxygen equilibrium of hemoglobin from Thunnus thynnus. Nature (London) 188: 895-896

Taylor MH, DiMichele L, Leach GJ (1977) Egg stranding in the life cycle of the mummichog, Fundulus heteroclitus. Copeia 1977: 397-399

Tsuyuki H, Roberts E, Vanstone WE (1965) Multiple hemoglobins of some members of the Salmonidae family. J. Fish Res. Board Can. 22: 203-213

Watt KWK, Riggs A (1975) Hemoglobins of the tadplole of the bullfrog Rana catesbliana. J. Biol. Chem. 250: 5934-5944

Weber RE (1975) Respiratory properties of hemoglobin from eunicid polychaetes. J. Comp. Physiol. 99: 297-307

Weber RE, Sullivan B, Bonaventura J, Bonaventura C (1976) The hemoglobin system of the primative fish Amia calva: Isolation and functional charaterization of the individual components. Biochim. Biophys. Acta 434: 18-31

Wittenberg JB, Wittenberg BA (1974) The choroid rete merabile of the fish eye. I. Oxygen secretion and structure: comparison with the swim bladder, rete mirabile. Biol. Bull. 146: 116-136

Wood SC, Johansen K (1972) Adapatation to hypoxia by increased HbO_2 affinity and decreased red cell ATP concentration. Nature 237: 278-279

Wood SC, Johansen K, Weber RE (1975) Effects of ambient pO_2 on O_2-Hb affinity and red cell ATP concentration in a benthic fish, Pleuronectes platessa. Resp. Physiol. 25: 259-267

Wyman J (1948) Heme proteins. Adv. Protein Chem. 19: 407-531

Wyman J (1964) Linked functions and reciprocal effects in hemoglobin: a second look. In: Anfinsen CB, Anson ML, Edsal JT, Richards FM (eds) Advances in Protein Chemistry, Vol. 19, Academic Press, New York, pp. 223-286

Wyman J, Gill SJ, Noll L, Giardina B, Colosima A, Brunori M (1977) The balance sheet of a hemoglobin: thermodynamics of CO binding by hemoglobin trout I. J. Mol. Biol. 109: 195-205

Yamaguchi K, Kochiyama Y, Matsurra F (1962) Studies on multiple hemoglobins of eel. II. Oxygen dissociation curve and relative amounts of components. F and S Bull. Japan Soc. Sci. Fish. 28: 192-198

Yamaguchi K, Kochiyama Y, Hashimoto K, Matsurra F (1963) Studies on two hemoglobins of loach. II. Oxygen dissociation curve. Bull. Japan Soc. Sci. Fish. 29: 180-188

Yamanaka H, Yamaguchi K, Matsurra F (1965) Starch gel electrophoresis of fish hemoglobins. I. Usefulness of cyanmethemoglobin for the electrophoresis. Bull. Japan Soc. Sci. Fish. 31: 827-832

Oxygen Transport Proteins: A Unitarian View Based on Thermodynamic Kinetic and Stereochemical Considerations

B. GIARDINA, M. COLETTA, L. ZOLLA, M. BRUNORI

I. INTRODUCTION

The metabolic needs of peripheral tissues are such that organisms have been forced, in the evolutionary sense, to develop special systems to transport oxygen from the outer environment to the tissues.

The chemical basis for O_2 transport is represented by the so-called respiratory proteins namely hemoglobins, erythrocruorins (i.e. invertebrate giant hemoglobins), hemocyanins and hemerythrins which differ greatly in the nature of the prosthetic group and of the protein moiety (Brunori et al., 1979; Brunori et al., 1982a) (Table 1).

TABLE 1. General Characteristics of Various Oxygen Carriers

Proteins	Source	Location	Prosthetic group	M.W.	N° of binding site
Tetrameric Hemoglobins	vertebrates	intracellular	heme	~ 65 000	4
Giant Hemoglobins	annelids molluscs arthropods	extracellular	heme	$> 10^6$	≥ 70
Hemocyanins	arthropods	extracellular	a pair of Cu(I)	$\geq 450\ 000$	≥ 6
	molluscs			$\geq 5 \times 10^6$	≥ 100
Hemerythrins	sipunculidis priapulidis brachiopods	intracellular	a pair of Fe(II)	~ 110 000	8

In spite of the very large chemical differences, the various O_2 carriers share a number of common basic features in relation to structural aspects and the molecular mechanisms underlying the fine tuning of their function.

In this note we outline some of these similarities in an attempt to provide a unified view of the basic features controlling the reactivity and the overall behaviour of all oxygen carriers. Indeed, basic homogeneity of behaviour emerges from a comparison of the available equilibrium and kinetic data analyzed within the framework of the two-state allosteric model (Monod et al., 1965) although suitable modifications involving essentially the definition of the so-called "functional constellations" (Colosimo et al., 1974) are rendered necessary.

Moreover, at the molecular level, similarities in function seem to find a common structural basis in the stereochemical perturbation, generally observed at the active site upon binding of oxygen, that very likely represents an initial event in the onset of the quaternary changes underlying the allosteric phenomena.

II. LIGAND BINDING AND STEREOCHEMICAL EFFECTS

Comparison of molecular events occurring on ligand binding in the various oxygen carrying proteins is not an easy task since the structural information available is at a very different degree of resolution, being very detailed for hemoglobin and still rather limited for the other macromolecules. However, on the basis of the extensive knowledge obtained from hemoglobins, some correlations may be proposed.

In the case of hemoproteins the upshot of the crystallographic results obtained both on myoglobin and hemoglobin (Baldwin and Chothia, 1979; Baldwin, 1980; Phillips, 1978; Shaanan, 1982) is that ligand binding brings about a tertiary perturbation of the active site, involving especially the interactions of the proximal histidine with the porphyrin.

Although the molecular mechanism underlying the energy flux from one heme to the others is still a matter for investigation, it is well proven that the structural perturbation, induced by ligand binding to the heme-iron, extends to the subunit interfaces altering the equilibrium between the two quaternary structures represented by the deoxy-and the oxy-forms of the molecule (Ackers, 1980; Bolton and Perutz, 1970). Thus, on the basis of the two-state model, the difference in free energy of binding between the two allosteric structures T (deoxy-) and R (oxy-) of the molecule can be associated with a tension developed at the proximal side of the heme-pocket whenever the ligands binds to the low-affinity T structure and which disappears as the quaternary switch takes place (Perutz, 1970).

Hemocyanins also display local changes in structure upon binding of oxygen as it has been shown by EXAFS, Raman and IR observation (Brown et al., 1980; van Holde and Miller, 1982).

The observed structural changes in going from deoxy- to oxyhemocyanin seem to be related to the necessity to optimize the copper to copper distance to allow a best fit of dioxygen. Thus, oxygen is bound as a bridging ligand to both copper atoms in a site and the structure of the metal-dioxygen-metal complex is that of a coplanar $Cu_2(II)$ peroxo. The copper-copper distance in deoxyhemocyanin is different from that of the oxygenated complex and at the same time the number of metal-coordinated internal ligands differs in the two quaternary conformations (Brown et al., 1980) (Fig. 1).

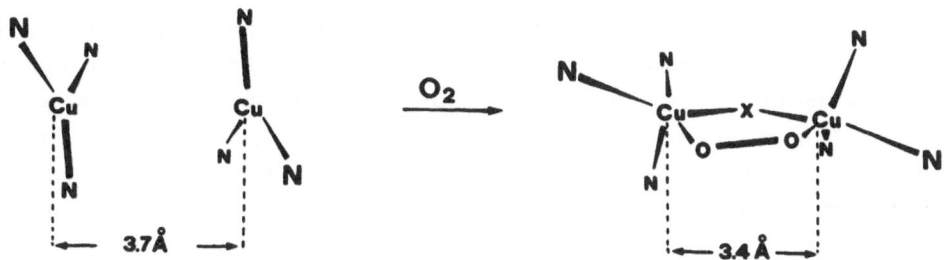

FIG. 1. A model for the structural change at the binding site of hemocyanin. In the deoxy-state, both Cu(I) have a trigonal planar arrangement being bound only to the three imidazole residues. Upon binding of O_2 both copper ions Cu(II) have a square-pyramidal coordination group whose basal plane is defined by the bridging dioxygen molecule, the internal oxygen atom (X) and two of the nitrogen atoms of imidazole. (After Brown et al., 1980).

Furthermore, at the level of the quaternary structure (Fig. 1), changes between oxy- and deoxyhemocyanin are also evident as indicated by ligand-linked subunit dissociation, proteolytic attack, optical probes and electron microscopy (van Holde and Miller, 1982). However, the lack of crystallographic information at the necessary level of resolution makes the identification of the structural changes almost impossible. In any case, the local perturbation required to fit dioxygen in between the two metals of hemocyanin may well represent an initial and essential event for the alteration of the quaternary equilibrium which in turn is a necessary prerequisite for the expression of cooperative phenomena (i.e. homotropic interactions).

In this respect it is interesting to point out that binding of carbon monoxide to hemocyanin shows no significant cooperativity. This marked difference between O_2 and CO may find a structural basis in the different modes of binding at the active site. Infrared studies performed using $^{13}CO^{18}$ (Fager and Alben, 1972; van der Deen and Hoving, 1979) have shown that carbon monoxide is bound to only one of the copper atoms, very likely through the oxygen with a CO/Cu stoichiometry of 1:2, and without interactions with the other copper atom in a site.

Cooperative effects should therefore be related to the bridging features that are unique to O_2, which acts as a "clamp" between the two copper atoms. This proposal is substantiated by the experiments showing that carbon monoxide binding is not associated with a quaternary transition (Brunori et al., 1982). Along this line it is suggestive that in hemerythrin, which is also a binuclear carrier using Fe^{2+} as complexing metal, dioxygen is bound in an end-on fashion at the active site like CO in Mb and homotropic interaction phenomena are not observed in spite of its polymeric nature (Loehr and Loehr, 1979).

III. COOPERATIVITY IN OXYGEN BINDING

The differences in the cooperative character of O_2 binding displayed by tetrameric hemoglobins (from vertebrates), giant hemoglobins (erythrocruorins from some invertebrates) and hemocyanins are quantitative rather than qualitative. Thus, in the case of extremely large respiratory proteins (both hemocyanins and erythrocruorins) which may contain even more than 100 oxygen binding sites, a very large value of the Hill coefficient (n_{50}: the slope of the Hill plot at 50% saturation) is often coupled with a relatively small total interaction free energy (ΔF_i) determined by the spacing of the asymptotes of the binding curve. In other words, the Hill plot shows a rather sharp upward bend, often in the middle range of saturation, associated with rather closely spaced asymptotes. The sharpness of the binding curve in relation to the spacing of the asymptotes (i.e. n_{50} vs ΔF_i) is an indication of the highly cooperative character of the transition between the functionally relevant states of the protein (Fig. 2). Such a behaviour is consistent also with predictions based on a two state model applied to a large system since in this case the features of the conformational transition, because of the increased number of binding sites, recall a phase change (Colosimo et al., 1974; Wyman, 1969). However analysis of data has shown that the number of actually interacting binding sites is much smaller than the total number of sites carried by the molecule. Thus applicability of the M.W.C. two-state model to giant respiratory proteins ruled out unequivocally the possibilities that homotropic interactions involve the whole molecule and demanded the introduction of an additional concept which implies that cooperativity is confined to subgroups of interacting sites called "functional constellations" (Colosimo et al., 1974). Following this model, a functional constellation is constituted only by a small fraction of the total number of sites on the macromolecule and represents a set of strongly interacting sites (presumably contiguous) within which most of the cooperative homotropic interactions are operative.

This description suggests that the variation of the interaction energy, induced at subunit interfaces by the tertiary structural changes which follow the ligand binding, does not spread out over the whole macromolecule but it is restricted to a limited number of subunits which act in a concerted fashion as a functional quasi-independent domain. The macromolecule can be therefore considered as a multi-domain structure where each functional constellation behaves almost independently of the others. However, some interactions among different constellations cannot be ruled out since interactions at their boundaries may well be energetically small but probably not zero.

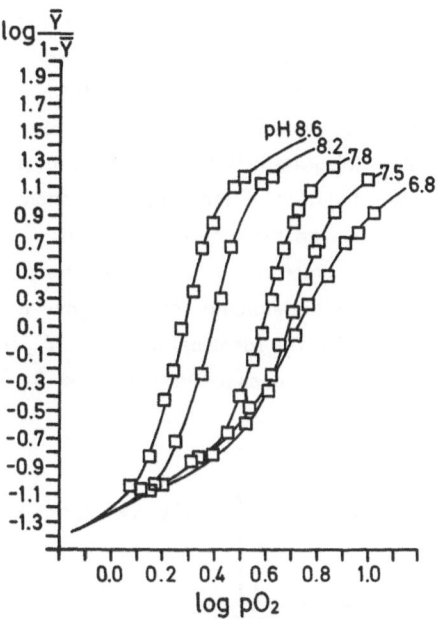

FIG. 2. Hill plots of native Octolasium complanatum erythrocruorin as a function of pH. The curves are theoretical ones computed on the basis of the allosteric model as described by Colosimo et al. (1974). The minimum number of interacting sites (functional constellation), within the pH range 8.6 to 7.5, is 12 that is the number of hemes carried by the 1/12 subunits. This number drops to 6 at pH 6.8.
Tris-HCl buffer, I = 0.1 M and temperature 20 °C (after Santucci et al., 1984).

In spite of this approximation the model provides a consistent description of the equilibrium data and allows a uniform treatment of results obtained on erythrocruorins and hemocyanins from different species and with different molecular weights.

Conceptually a similar description can be applied also to tetrameric hemoglobins, namely hemoglobin A, even though in this case the structural basis of cooperativity is known in greater details allowing to correlate the energetics of the process to specific interactions within the tetramer (Baldwin, 1975; Perutz, 1970). Hence, cooperativity stems from the fact that, in the absence of ligand, the T form is energetically more stable than the R form and binding of oxygen to the T-state is less favorable than binding to the R-state by about 3 Kcal/site. This estimate corresponds to a free energy of interactions (ΔF_i) of 12 Kcal per mole of tetramer. A substantial amount of work has been done, in the last few years, in order to elucidate the localization of this difference in free energy of O_2 binding (between the two extreme forms of the molecule), in which way the tension of the bound T-state is released upon the quaternary switch as well as the energetics of the subunit interactions (Baldwin and Chothia, 1979; Perutz, 1972; Valdes and Ackers, 1977 and 1978). From these studies it turns out as a definitive piece of information that the bond between the metal atom and the proximal histidine is involved in the spreading of the conformational change from the metal-ligand complex to the globin. However, it seems clear that the amount of energy "stored" at the level of the heme is small (0.3 to 1 Kcal per site, according to some authors) and that a large fraction should be stored elsewhere in the molecule, very likely at the subunit interfaces (Nagai and Kitagawa, 1980). In this connection studies on the relationships between ligand binding and stability of monomers, dimers and tetramers (Ackers, 1980) appear to indicate an asymmetric distribution of energy between the two possible dimers into which a hemoglobin molecule may dissociate, i.e. $\alpha_1 \beta_2$ (or $\alpha_2 \beta_1$) and $\alpha_1 \beta_1$ (or $\alpha_2 \beta_2$). Thus the hemoglobin tetramer may be thought to be formed by association of the $\alpha_1 \beta_1$ dimer with the identical dimer $\alpha_2 \beta_2$ leading to contacts between α_1 and β_2 as well as the symmetrical ones between α_2 and β_1 (Fig. 3).

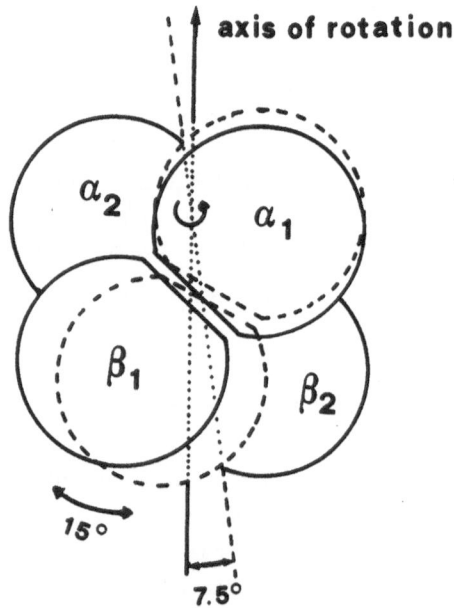

FIG. 3. Schematic diagram of quaternary change in human hemoglobin A. Upon transition from the T (continuous lines) to the R structure (dashed lines) the $\alpha_1\beta_1$ dimer rotates 15° relative to $\alpha_2\beta_2$ about the axis of symmetry which is rotated 7.5 degrees. $\alpha_1\beta_1$ dimer moves as a rigid body so that one type of contact (labeled $\alpha_2\beta_2$ or $\alpha_1\beta_1$) is identical in the two forms while the other (labeled $\alpha_1\beta_2$ or $\alpha_2\beta_1$) differs. (After Baldwin, 1980).

In addition, it has been observed very recently (Ackers, G., personal communication) that in an asymmetric hybrid tetramer from HbA and Hb Kempsey ($\alpha_1^h \ \alpha_2^k \ \beta_1^h \ \beta_2^k$), where the interfaces $\alpha_1 \beta_2$ and $\alpha_2 \beta_1$ have different intersubunit contacts such that in the deoxy-form, only one interface is able to assume a T-like structure, the energetic difference between the unliganded and liganded forms is approximately half of that of normal human hemoglobin A.

This would suggest that cooperativity is mostly transmitted through these interfaces and that information for this structural change is transferred only to a negligible extent through other intersubunit contacts such as $\alpha_1\beta_1$, $\alpha_2\beta_2$, and $\alpha_1\alpha_2$. This view is consistent with the analysis of tridimensional structures of deoxy and liganded hemoglobin (Perutz and Ten Eyck, 1971) showing that very relevant structural changes occur at the $\alpha_1\beta_2$ interface where a few amino acid side chains seem to acquire critical positions in the two different allosteric states of the molecule. This would also account for the well established absence of cooperative phenomena within the free $\alpha_1\beta_1$ dimer (Hewitt et al., 1972) whose interface does not undergo any structural change on going from oxy- to deoxyform (Baldwin, 1980).

All these considerations lead back to the so-called rectangular model proposed many years ago (Wyman, 1948). The model may be depicted as an assembled tetramer with much stronger <u>functional</u> interactions within the $\alpha_1\beta_2$ (and $\alpha_2\beta_1$) dimer and much weaker between $\alpha_1\beta_1$ (and $\alpha_2\beta_2$) dimer.

Hence, following the model developed for giant oxygen carriers within a tetramer of HbA, we may distinguish two "functional constellations" represented by the $\alpha_1\beta_2$ and $\alpha_2\beta_1$ dimers only weakly but definitely interacting with each other as demanded by $n_{50} > 2$.

A further point of interest concerns the possible relationship between the above defined functional constellations and the structural domains which have been identified by a series of investigations of the structural properties of hemocyanins based on electron microscopy (Siezen et al., 1974) and limited proteolysis studies (Brouwer, 1975; Gielens et al., 1981). In this connection it should be remarked that the two different kind of dimers present in HbA (i.e. $\alpha_1\beta_2$ and $\alpha_1\beta_1$) are characterized by different stabilization energies since dissociation of the tetrameric molecule occurs through a preferential cleavage of one type of interface leading to non cooperative dimers ($\alpha_1\beta_1$ or $\alpha_2\beta_2$) with a rupture of the functionally relevant contacts. This also seems a general phenomenon since in almost all O_2 carriers dissociation into subunits of lower molecular weight is accompanied by the loss or a strong reduction of cooperativity (Antonini and Chiancone, 1977; Brunori et al., 1984a; van Holde and Miller, 1982). It looks as though the intersubunit contacts which are preferentially disrupted by dissociation are those which are also primarily involved in the expression of cooperativity and hence in the maintenance of an operative functional constellation.

Within this description it is interesting to point out that hemoglobin behaviour parallels that of other oxygen carriers since it can be considered as a macromolecule composed by two dimeric functional constellations but which are operative only in the fully associated form, as observed in most of the respiratory proteins investigated up to now.

IV. HETEROTROPIC INTERACTIONS: THE BOHR EFFECT

The position of the oxygenation curve (determined by p_{50}, i.e. the O_2 pressure corresponding to 50% saturation) in relation to oxygen pressure is of critical physiological importance, since it determines the amount of O_2 released at level of tissues. A number of factors, such as temperature and neutral salt, can affect p_{50} thereby acting as modulators of hemoglobin function.

The first recognized control mechanism for oxygen transport, known as the "Bohr effect" represents the heterotropic linkage between O_2 and proton binding sites and finds its physical basis in the ligand-linked conformational transition which affects the ionization equilibria of amino acid side chains (Perutz, 1970 and 1972). The identification of specific amino acid residues of the α and β chains of human hemoglobin A involved in this effect is complicated by differential interactions of other charged solution components (anions and cations) to deoxy- and oxyhemoglobin. Thus a preferential binding of a given anion (i.e. Cl^- and/or 2,3 DPG) to deoxyhemoglobin involves the alteration of the pK of some cationic groups, thereby contributing to the overall observed Bohr effect. However, studies on hemoglobins from various species and of several abnormal hemoglobins (Perutz et al., 1980) have allowed to identify some of these residues; those involved in the alkaline Bohr effect are reported in Table 2 that indicates also if a particular ionizable group is contributing to the Bohr effect through a differential binding of Cl^- and/or 2,3 DPG. Apart from the specific residues involved, in accord with the two-state model the effect of protons on the binding of oxygen to hemoglobin has to be correlated to their different affinity towards the two different quaternary conformations. That this is the case is demonstrated by a substantial difference observed in the number of H^+ ions that are bound to the protein in the ligand-free and ligand-bound forms (Antonini and Brunori, 1971). However, on the basis of accurate equilibrium measurements (Imai and Yonetani, 1975) it has been suggested that in human hemoglobin A the Bohr effect is a mixture of tertiary and quaternary contributions.

A definite correlation between the Bohr effect and the quaternary conformational change has been established on the basis of crystallographic data (Anderson, 1975) and has been furtherly supported by the absence of heterotropic interactions in the component I of the trout Hb system which is characterized by substitution of residues (Barra et al., 1981) previously proposed to play an important role in modulating the influence of non-heme ligands on the functional properties of human hemoglobin (Perutz, 1970) (see Table 2).

In addition, it is known that some fish hemoglobins are characterized by a very marked Bohr effect (called Root effect) so that a decrease in pH lowers their oxygen affinity to such an extent that, for pH values < 7, the protein is only partially saturated with O_2 even at atmospheric pressure. This marked decrease in O_2 affinity, related from the physiological stand point both to the presence and function of the swimbladder (Scholander and van Dam, 1954) and to O_2 secretion in the eye, is accompanied by a very large effect of

TABLE 2. Groups of the Alkaline Bohr Effect of Human Hemoglobin A
In comparison the amino acid residues occupying the same positions in the component I of
trout Salmo irideus are reported (after Brunori et al., 1982)

Hb A	2,3 DPG	Chloride	Trout Hb I
α 1(NA2)Val	no	yes	Ac-Ser
α 122(H5)His	no	?	His
β 1(NA1)Val	yes	±	Val
β 2(NA2)His	yes	±	Glu
β 143(H21)His	yes	±	Ser
β 146(HC3)His	no	no	Phe
β 82(EF6)Lys	yes	±	Leu

pH on the shape of the ligand binding curve which becomes less and less cooperative as the
pH is lowered tending to values of the Hill coefficient, n_{50}, equal to 1 or even lower. This
particular behaviour has been explained (leaving aside the role of chain functional
heterogeneity) on the basis of a marked stabilization of the T-conformation induced by
protons which dramatically alters the relative stability of the two allosteric states (Brunori
et al., 1978; Noble et al., 1970; Wyman et al., 1978).

A stereochemical interpretation of the Root effect has been recently proposed on the
basis of considerations of the tridimensional model of HbA and sequence information on fish
and amphibian hemoglobins (Perutz and Brunori, 1982). Thus, the presence of a serine
residue at postion F9 (β 93), which in hemoglobins characterized by the Root effect
substitutes for the cysteinyl residue normally found in mammals, allows a stabilization of
the T-quaternary state at low pH values (say about 6.5) because of the formation of two
additional H-bonds when His HC3 (β 146) is protonated. The formation of these bonds with
the protonated form of His HC3 (β 146) leads to the pH dependence of the allosteric
equilibrium constant with a predominance of the T-state at low pH in both the unliganded
and liganded forms (Fig. 4). Further support for this stereochemical explanation of the Root
effect comes from a recent kinetic investigation of the functional properties of the
hemoglobin from the tadpole (stage 55) of Xenopus laevis which is known to possess a
phenylalanine and an alanine residue at position 146β and 93β , respectively. In this case in
fact, the kinetic analysis has clearly demonstrated the disappearance of the Root effect
(Brunori et al., 1984b).

A similar type of phenomenon which leads to the stabilization of one of the two
quaternary conformations (either T or R) is often observed with erythrocruorins and
hemocyanins where it has been shown that not only protons, but anions and cations as well,
may dramatically affect the shape of the oxygen binding curve (Brouwer et al., 1978; Kuiper

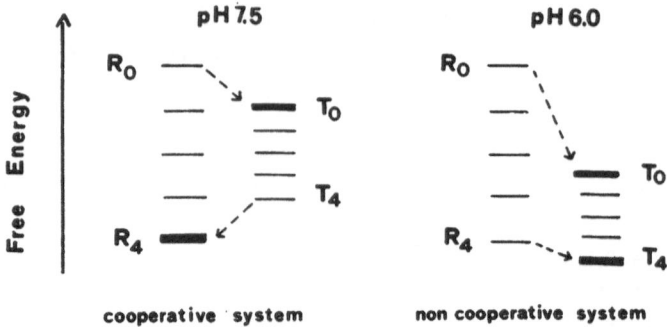

FIG. 4. Energy levels diagram for a "Root effect" hemoglobin. At alkaline pH, where the system is cooperative, the fully unliganded protein (sub o) is stable in the T state whereas the saturated molecule (sub 4) is stabilized in the R state. At acid pH, the T state is more stable independently of the degree of saturation and therefore the system is non-cooperative. (After Brunori et al., 1984).

et al., 1979; Santucci et al., 1984; Zolla et al., 1978) (Fig. 5). As it is evident from figure 5 in the case of β-hemocyanin from Helix pomatia, pH values higher than 7.2 bring about a progressive stabilization of the T-state of the molecule while an opposite stabilization of the R-state occurs on the other side (pH < 7.2).

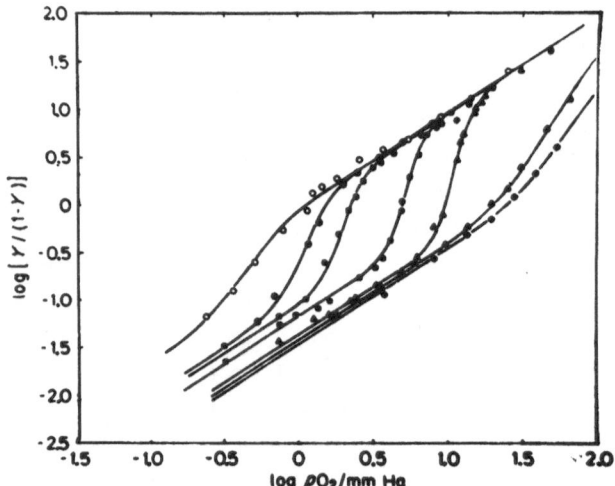

FIG. 5. Hill plots of the O_2 equilibrium curves of β-hemocyanin from Helix pomatia. The continuous lines have been computed according to a two-state model as described by Colosimo et al. (1974).
The minimum number of interacting sites necessary to account for the observed cooperativity is 15 (pH range 7-7.4).
Conditions: 0.02 M Tris-HCl + 17 mM Ca^{2+}
(○): pH 7; (●): pH 7.05; (□): pH 7.1; (■): pH 7.2; (△): pH 7.4; (▲): pH 8.0; (✷): pH 9.0; (after Zolla et al., 1978).

However, the low affinity T-state (lower asymptote in the Hill plot) is clearly pH dependent contrary to what expected on the basis of an original two-state model for which an allosteric effector is supposed to change exclusively the ratio of the quaternary states. Hence, although the main part of the regulation comes from changes in the allosteric constant, the behaviour of β-hemocyanin from H. pomatia seems to demand the existence of a set of T-states whose equilibrium is influenced by pH.

Observation on other hemocyanins (Brouwer et al., 1977; Er-el et al., 1972) and erythrocruorins (Santucci et al., 1984; Weber, 1981) show that modifications of the T-state (or the R-state) by an allosteric effector may be a general phenomenon. In this respect, it is worth recalling that a similar situation occurs also in human hemoglobin since the lower asymptote of the binding curve appears significantly influenced by protons and 2,3 DPG (Imai, 1973; Imai and Yonetani, 1975).

A general picture which emerges is that an allosteric effector, besides shifting the equilibrium between the functionally relevant states of a protein, in many cases also influences the properties of that state which is favourably stabilized by the effector.

This general behaviour emphasises the overall similarities among different O_2 carriers since heterotropic effectors display in all of them a similar pattern of phenomena which can be interpreted, at least to a first approximation, within the framework of the two-state M.W.C. model.

V. KINETICS OF OXYGEN BINDING

Kinetics of oxygen binding has been characterized extensively for hemoglobins and hemocyanins from various species (Brunori et al., 1982a; Parkhurst, 1979).

It should be outlined that the quaternary switch is generally very fast compared with binding or dissociation of oxygen, at least for those proteins which are characterized by a small functional constellation ($\leq$ six sites) (Hopfield et al., 1971). Hence, in hemoglobin and in hexameric hemocyanins the relaxation time for this process has been estimated $\geq 10^4 s^{-1}$ (Sawicki and Gibson, 1977). Moreover, a relationship between the overall size of the functional constellation and the speed of the quaternary conformational change has been clearly established (van Driel et al., 1978); thus, when the size of the allosteric unit becomes larger, the quaternary conformational change becomes slower and in some cases it appears to be rate limiting (Kuiper et al., 1978; van Driel et al., 1974).

In any case, the kinetics of the R and T states can be characterized independently by working in a suitable saturation range that is at very high and very low oxygen saturations, respectively. Some of the representative rate constants for the reaction with O_2 of hemoglobins and hemocyanins in both quaternary states as well as the corresponding non-cooperative subunits are reported in Table 3. The set of data outlines some common features which may be summarized as follows:

i - The second order rate constant for O_2 binding is essentially independent of allosteric state being in all cases very large (10^7 - 10^8 $M^{-1}s^{-1}$) and tending towards diffusion controlled.

ii - The first order dissociation rate constant is strongly dependent on quaternary structure and represents the major source of difference in the binding energy between R and T.

TABLE 3. Association (k') and Dissociation (k) Kinetic Constants for the Reaction with Oxygen of Some Respiratory Proteins (after Brunori et al., 1984, modified)

Protein	$k'_T(M^{-1}s^{-1})$	$k'_R(M^{-1}s^{-1})$	$k_T(s^{-1})$	$k_R(s^{-1})$
α -chain (from HbA)	-	50	-	20
β -chain (from HbA)	-	71	-	17
Mb (sperm-whale)	-	19	-	11
α -chain (within the HbA tetramer)	2.9	59	183	12
β -chain (within the HbA tetramer)	12	59	2500	21
β -Hemocyanin (from Helix pomatia)	5	5	700	5
Hemocyanin hexamer (from Panulirus interruptus)	-	31	-	60
Hemocyanin monomer (from Panulirus interruptus)	46	-	1500	-
Erythrocruorin (from Lumbricus terrestris)	-	30	-	60

In conclusion, the main kinetic regulatory mechanism of the cooperativity at the active site seems to lie in the different tendency of the oxygen-metal complex to dissociate as between the two conformational states. This confirms the general rule that, from a kinetic viewpoint, the process of control is exerted mainly on the "off" rates.

VI. EVOLUTIONARY CONSIDERATIONS

The reversible combination with molecular oxygen is obviously the basic chemical property of all respiratory proteins, being the main requirement of any oxygen carrier. However, oxygen carriers provide a good example of different chemical strategies which have been adopted during evolution to achieve a consistent solution to the vital problem of O_2 supply to tissues for metabolic demands. In this context, it is clear that homotropic and heterotropic effects, as well as the kinetic basis of cooperative O_2 binding, are common and can be described taking into account ligand-linked conformational changes which may have a similar stereochemical basis.

Together with these similarities in function, the distribution of the different respiratory pigments among the various organisms offers some suggestive considerations with respect to the appearance of oxygen transporters during the course of evolution. Therefore, it should be recalled that among molluscs, for example, the bivalves have hemoglobins and the cephalopods have hemocyanins and that sometimes the same individual has the capability of producing both types of oxygen carriers as demonstrated by several species displaying hemocyanin in the hemolymph and myoglobin in their muscles (as exemplified by Aplysia limacina).

Given all of this, we are confronted with the interpretation of the mechanism by which two chemically distinct O_2 carriers appear in the same phylum and/or in the same species.

Although an unequivocal solution of this problem is not possible at the moment, a likely hypothesis could be the development, during evolution, of two independent chemical species which have both been used as oxygen transporters and which only thereafter have been "selected" for, so that nowadays some species use hemocyanins and other hemoglobins. If this were the case it might be possible that in several species the information for synthesis of both types of O_2 carrier is still present but selectively repressed.

REFERENCES

Ackers GK (1980) Energetics of subunit assembly and ligand binding in human haemoglobin. Biophys. J. 32: 331-346

Anderson L (1975) Structures of deoxy and CO haemoglobin Kansas in the deoxy quaternary conformation. J. Mol. Biol. 94: 33-49

Antonini E, Brunori M (1971) Haemoglobin and myoglobin in their reaction with ligands. Elsevier, North-Holland Amsterdam

Antonini E, Chiancone E (1977) Assembly of multisubunit respiratory proteins. Ann. Rev. Biophys. Bioeng. 6: 239-271

Baldwin JM (1975) Structure and function of haemoglobin. Progr. Biophys. Mol. Biol. 29: 225-320

Baldwin JM (1980) The structure of human carbon monoxy haemoglobin at 2.7 Å resolution. J. Mol. Biol. 136: 103-128

Baldwin JM, Chothia C (1979) Haemoglobin: the structural change related to ligand binding and its allosteric mechanism. J. Mol. Biol. 129: 175-187

Barra D, Bossa F, Brunori M (1981) Structure of binding site for heterotropic effectors in fish haemoglobins. Nature (London) 228: 587-588

Bolton W, Perutz MF (1970) Three dimensional Fourier synthesis of horse deoxyhaemoglobin at 2.8 Å resolution. Nature (London) 228: 551-552

Brouwer M (1975) Structural domains in Helix pomatia α-haemocyanin. Doctoral dissertation, Rijksuniversiteit, Groningen

Brouwer M, Bonaventura C, Bonaventura J (1977) Oxygen binding by Limulus polyphemus haemocyanin: allosteric modulation by chloride ions. Biochemistry 16: 3897-3902

Brouwer M, Bonaventura C, Bonaventura J (1978) Analysis of the effect of three different allosteric ligands on O_2 binding by haemocyanin of the shrimp Penaeus setiferus. Biochemistry 17: 2148-2154

Brown JM, Powers L, Kincaid B, Larrabee JA, Spiro TG (1980) Structural studies of the hemocyanin active site. Extended X-ray absorption fine structure (EXAFS) analysis. J. Amer. Chem. Soc. 102: 4210-4216

Brunori M (1975) Molecular adaptation to physiological requirements: the haemoglobin system of trout. Curr. Topics Cell. Reg. 9: 1-39

Brunori M, Coletta M, Giardina B, Wyman J (1978) A macromolecular transducer as illustrated by trout Hb IV. Proc. Natl. Acad. Sci. USA 75: 4310-4312

Brunori M, Giardina B, Bannister JV (1979) Oxygen-transport proteins. In: H.A.O. Hill (ed) Inorganic Biochemistry, Vol. I, Royal Society of Chemistry, London, pp. 159-209

Brunori M, Giardina B, Kuiper HA (1982a) Oxygen-transport proteins. In: H.A.O. Hill (ed) Inorganic Biochemistry, Vol. III, Royal Society of Chemistry, London, pp. 126-182

Brunori M, Kuiper HA, Zolla L (1982b) Ligand binding and stereochemical effects in haemocyanin. EMBO J. 1: 329-331

Brunori M, Coletta M, Giardina B (1984a) Oxygen Carrier Proteins. In: Harrison P (ed) Metalloproteins, The MacMillan Press Ltd (in press)

Brunori M, Condò SG, Belleli A, Giardina B, Micheli G (1984b) Tadpole Xenopus laevis haemoglobin: correlation between structure and functional properties. J. Mol. Biol. (submitted)

Brunori M, Giacometti GM, Giardina B (1984c) The reaction of haemoglobin with oxygen. In: Bannister JV, Bannister WH (eds) The Biology and Chemistry of Active Oxygen, Elsevier, Amsterdam, pp. 16-44

Colosimo A, Brunori M, Wyman J (1974) Concerted changes in an allosteric macromolecule. Biophys. Chem. 2: 338-344

Er-el Z, Shaklai N, Daniel E (1972) Oxygen binding properties of haemocyanin from Levantina hierosolima. J. Mol. Biol. 64: 341-352

Fager LY, Alben JO (1972) Structure of the CO binding site of haemocyanin studied by Fourier transform IR spectroscopy. Biochemistry 11: 4786-4792

Gielens C, Verschueren LJ, Préaux G, Lontie R (1981) Localization of the domains in the polypeptide chain of β_c-haemocyanin of Helix pomatia. In: Lamy J, Lamy J (eds) Invertebrate Oxygen Binding Proteins, Marcel Dekker, New York, pp. 295-304

Hewitt JA, Kilmartin JV, Ten Eyck LF, Perutz MF (1972) Non-cooperative of the α β dimer in the reaction of haemoglobin with oxygen. Proc. Natl. Acad. Sci. USA 69: 203-207

Hopfield JJ, Shulman AG, Ogawa S (1971) An allosteric model of hemoglobin: Kinetics. J. Mol. Biol. 61: 425-443

Imai K (1973) Analyses of O_2 equilibria of native and chemically modified human haemoglobin on the basis of Adair's stepwise oxygenation theory and the allosteric model of Monod, Wyman and Changeux. Biochemistry 12: 798-807

Imai K, Yonetani T (1975) pH dependence of the Adair constants of human haemoglobin. J. Biol. Chem. 250: 2227-2231

Kuiper HA, Brunori M, Antonini E (1978) Kinetics of the Bohr effect in the reaction of Helix pomatia β -haemocyanin with oxygen. Biochem. Biophys. Res. Commun. 82: 1062-1069

Kuiper HA, Forlani L, Chiancone E, Antonini E, Brunori M, Wyman J (1979) Multiple linkage in Panulirus interruptus haemocyanin. Biochemistry 18: 5849-5854

Loehr JS, Loehr TM (1979) Haemerythrin: a review of structural and spectroscopic properties. In: Eichhorn GL, Marzilli LO (eds) Advances in Inorganic Biochemistry, Vol. I, Elsevier North-Holland, Amsterdam, pp. 235-253

Monod J, Wyman J, Changeux JP (1965) On the nature of allosteric transitions: a plausible model. J. Mol. Biol. 12: 88-118

Nagai K, Kitagawa T (1980) Differences in Fe(II)-N (His F8) stretching frequencies between deoxy-haemoglobin in the two alternative quaternary structures. Proc. Natl. Acad. Sci. USA 77: 2033-2037

Noble RW, Parkhurst LJ, Gibson QH (1970) The effect of pH on the reactions of O_2 and CO with the haemoglobin of the carp Cyprinus carpio. J. Biol. Chem. 245: 6628-6633

Parkhurst LJ (1979) Haemoglobin and myoglobin ligand kinetics. Ann. Rev. Phys. Chem. 30: 503-546

Perutz MF (1970) Stereochemistry of cooperative effects in haemoglobin. Nature (London) 228: 726-734

Perutz MF (1972) Nature of haem-haem interactions. Nature (London) 237: 495-499

Perutz MF, Ten Eyck LF (1971) Stereochemistry of cooperative effects in haemoglobin. Symp. Quant. Biol. XXXVI: 295-309

Perutz MF, Brunori M (1982) Stereochemistry of cooperative effects in fish and amphibian haemoglobins. Nature (London) 299: 421-426

Perutz MF, Kilmartin JV, Nishikura K, Fogg JM, Butler PJG, Rollema HS (1980) Identification of residues contributing to the Bohr effect of human haemoglobin. J. Mol. Biol. 138: 649-670

Phillips SEV (1980) Structure of oxymyoglobin. Nature 273: 247-248

Phillips SEV (1980) Structure and refinement of oxymyoglobin at 1.6 Å resolution. J. Mol. Biol. 142: 531-554

Santucci R, Chiancone E, Giardina B (1984) Oxygen binding to Octolasium complanatum erythrocruorin: modulation of homo- and heterotropic interactions by cations. J. Mol. Biol. (In press)

Sawicki CA, Gibson QH (1977) Properties of the T-state of human oxyhaemoglobin studied by laser photolysis. J. Biol. Chem. 252: 7538-7547

Scholander PF, van Dam L (1954) Secretion of gases against high pressures in the swimbladder of deep sea fishes. Biol. Bull. 107: 247-259

Shaanan B (1982) The iron-O_2 bond in human oxyhaemoglobin. Nature (London) 296: 683-684

Siezen RJ, van Bruggen EFJ (1974) Structure and properties of haemocyanins. XII. Electron microscopy of dissociation products of Helix pomatia α-haemocyanin quaternary structure. J. Mol. Biol. 90: 77-89

Valdes R, Ackers GK (1977) Thermodynamic studies on subunit assembly in human haemoglobin. J. Biol. Chem. 252: 74-81

Valdes R, Ackers GK (1978) Self-association of haemoglobin β^{SH} chains is linked to oxygenation. Proc. Natl. Acad. Sci. USA 75: 311-314

van der Deen H, Hoving H (1979) An infrared study of CO complexes of haemocyanin: evidence for the structure of the CO binding site from vibrational analysis. Biophys. Chem. 9: 169-179

van Driel R, Brunori M, Antonini E (1974) Kinetics of the cooperative and non-cooperative reaction of Helix pomatia haemocyanin with oxygen. J. Mol. Biol. 89: 103-112

van Driel R, Kuiper HA, Antonini E, Brunori M (1978) Kinetics of the cooperative reaction of Helix pomatia haemocyanin with oxygen. J. Mol. Biol. 121: 431-439

van Holde KE, Miller KI (1982) Haemocyanins. Quart. Rev. Biophys. 15: 1-45

Weber RE (1981) Cationic control of O_2 affinity in lunworm erythrocruorin. Nature 292: 386-387

Wyman J (1948) Heme Proteins. Adv. Prot. Chem. 4: 407-531

Wyman J (1969) Possible allosteric effects in extended biological systems. J. Mol. Biol. 39: 523-538

Wyman J, Gill SJ, Gaud HT, Colosimo A, Giardina B, Kuiper HA, Brunori M (1978) Thermodynamics of ligand binding and allosteric transition in haemoglobin. Reaction of trout Hb IV with CO. J. Mol. Biol. 124: 161-175

Zolla L, Kuiper HA, Vecchini P, Antonini E, Brunori M (1978) Dissociation and oxygen binding behaviour of β-haemocyanin from Helix pomatia. Eur. J. Biochem. 87: 467-473

Evolution and Adaptation of
Avian and Crocodilian Hemoglobins

A.G. SCHNEK, C. PAUL, J. LEONIS

I. INTRODUCTION

Among the reptiles, the species belonging to the crocodilia order are the closest relatives of birds (Romer, 1966). Indeed the thecodontia order probably arose directly from the primitive cotylosaurs during the Triassic period some 220 millions years ago. These thecodonts gave rise to crocodiles during the Jurassic period and birds during the Triassic about 150 millions years ago.

It is well known that most vertebrates possess a regulatory mechanism induced by specific cofactors that leads to a decrease in blood oxygen affinity compared to the intrinsic value of the isolated hemoglobin.

As Perutz pointed out recently, it is now possible to distinguish some substitutions in the animal hemoglobins that seem to have been brought about by selection or even by adaptative changes (Perutz, 1983).

A thorough examination of the molecular structure of crocodilian and avian hemoglobins and a detailed investigation of their oxygenation mechanism support this theory. Crocodiles and birds, despite their nearness on the evolutionary tree, have evolved completely different strategies of allosterically controlled hemoglobin oxygenation.

II. CROCODILIAN HEMOGLOBINS

A. Oxygen Affinity

The oxygen affinity of isolated hemoglobin from various crocodile species is high compared to its value obtained inside the red blood cells (Bauer and Jelkmann, 1977). However, contrary to most vertebrate hemoglobins, crocodilian hemoglobins are not inhibited by the usual allosteric effectors, namely various phosphate esters, carbamino CO_2^- and chloride.

The first observations performed on crocodilian blood samples already suggested this behavior. Indeed, 2,3-diphosphoglycerate (2,3-DPG) was not enzymatically detectable in crocodilian erythrocytes and ATP was present at lower concentrations than hemoglobin (Rapoport and Guest, 1941). All this indicated that these cofactors could not play a functional role in the hemoglobin oxygenation. Also Dill and Edwards (1931) noticed the large influence of CO_2 on crocodilian red blood cell oxygenation.

Bauer and collaborators reinvestigated thoroughly this oxygenation mechanism. They showed that neither ATP, nor 2,3-DPG, nor inositol pentaphosphate (IP5) has any significant influence on the oxygen affinity of crocodile hemoglobin (Bauer and Jelkmann, 1977). Instead, an analogous allosteric effect is obtained by addition of CO_2 equivalent to a PCO_2 of 40 torr (5.3 kPa). At pH 7.4 and 20 °C, in 0.1 M NaCl, that PCO_2 increases P_{50} (pressure at which hemoglobin is half saturated) ninefold from 3.4 (0.45 kPa) to 30.0 torr (4 kPa) (Jelkmann and Bauer, 1980). They discussed the mechanism by which CO_2 modifies oxygen binding (Bauer et al., 1981). It is in the form of hydrogenocarbonate ion that CO_2 exerts its effect on hemoglobin oxygen affinity, two anions being bound per tetramer. The association constant depends on the state of ligation; caiman deoxyhemoglobin presents an association constant of $2.10^{-3} M^{-1}$ at pH 7 and 25 °C whereas the oxy form does not display any binding. Also CO_2 reduces the Hill coefficient to a value of 2.4, which means that the allosteric equilibrium is greatly shifted toward the tense (T) deoxy-structure, so keeping the oxygen binding from being fully cooperative.

We measured the kinetic constant for the deoxygenation of caiman hemoglobin by stopped flow method (Leclercq et al., 1980). Again, 2,3-DPG or inositol hexaphosphate (IP6) have no effect but there is an increase in this parameter from 26.3 sec^{-1} to 43.0 sec^{-1} induced by a PCO_2 of 76 torr (10.1 kPa) at pH 7.4 and 25 °C.

B. Amino acid sequences

In order to explain the particular effect of hydrogen carbonate ion at the molecular level, the primary structures of three crocodilian hemoglobins - those of the Nile (Crocodylus niloticus) Mississipi (Alligator mississipiensis) and caiman (Caiman crocodylus) were determined. This work was performed in the framework of a collaboration with the laboratory of Professor Braunitzer at the Max Planck Institute (Leclercq et al., 1981).

The amino acid sequence of the α and β chains of the three hemoglobins are aligned on figures 1 and 2 and are compared to the human chains. The residues implicated in the interactions between the subunits and in binding to the cofactor are indicated.

C. Relation between structure and activity

Compared to the human hemoglobin each crocodilian α chain displays about fifty and each β chain about seventy amino acid substitutions. Only a very few at all these are necessary to explain the difference in the behavior of those proteins namely their specific binding to one or another effector.

```
                        1                    10                       20                       30
                                                                                    π
Human     Val-Leu-Ser-Pro-Ala-Asp-Lys-Thr-Asn-Val-Lys-Ala-Ala-Trp-Gly-Lys-Val-Gly-Ala-His-Ala-Gly-Glu-Tyr-Gly-Ala-Glu-Ala-Leu-Glu-Arg-Met-
Nile Cr.  Val-Leu-Ser-Ser-Asp-Asp-Lys-Cys-Asn-Val-Lys-Ala-Val-Trp-Ser-Lys-Val-Ala-Gly-His-Leu-Glu-Glu-Tyr-Gly-Ala-Glu-Ala-Leu-Glu-Arg-Met-
Miss.All. Val-Leu-Ser-Met-Glu-Asp-Lys-Ser-Asn-Val-Lys-Ala-Ala-Trp-Gly-Lys-Val-Gly-Gly-His-Leu-Glu-Glu-Tyr-Gly-Ala-Glu-Ala-Leu-Glu-Arg-Met-
Caiman    Val-Leu-Ser-Val-Lys-Lys-Asp-Lys-Ile-Trp-Gly-Lys-Val-Ala-Gly-His-Leu-Glu-Glu-Tyr-Gly-Ala-Glu-Ala-Leu-Glu-Arg-Met-

                                       40  σ   σ                      50                       60   σ    σ    σ
                                       π   π   π   σ
Human     Phe-Leu-Ser-Phe-Pro-Thr-Thr-Lys-Thr-Tyr-Phe-Pro-His-Phe-Asp-Leu-Ser-His-Gly-Ser-Ala-Gln-Val-Lys-Gly-His-Gly-Lys-Lys-Val-Ala-Asp-
Nile Cr.  Phe-Ser-Ala-Tyr-Pro-Gln-Thr-Lys-Ile-Tyr-Phe-Pro-His-Phe-Asp-Leu-Ser-His-Gly-Ser-Ala-Gln-Ile-Lys-Gly-His-Gly-Lys-Lys-Val-Ala-Phe-Ala-
Miss.All. Phe-Cys-Ala-Tyr-Pro-Gln-Thr-Lys-Ile-Tyr-Phe-Pro-His-Phe-Asp-Met-Ser-His-Asn-Ser-Ala-Gln-Ile-Arg-Ala-His-Gly-Lys-Lys-Val-Val-Phe-Ser-
Caiman    Phe-Cys-Ala-Tyr-Pro-Gln-Thr-Lys-Ile-Tyr-Phe-Pro-His-Phe-Asp-Leu-Ser-His-Gly-Ser-Ala-Gln-Ile-Arg-Ala-His-Gly-Lys-Lys-Val-Phe-Ala-

                        70                       80                       90   σ    σ    σ    σ
Human     Ala-Leu-Thr-Asn-Ala-Val-Ala-His-Val-Asp-Asp-Met-Pro-Asn-Ala-Leu-Ser-Ala-Leu-Ser-Asp-Leu-His-Ala-His-Lys-Leu-Arg-Val-Asp-Pro-Val-
Nile Cr.  Ala-Leu-His-Glu-Ala-Val-Asn-His-Ile-Asp-Asp-Ile-Ala-Leu-Pro-Asp-Leu-Ser-Arg-Leu-His-Ala-His-Ser-Leu-Arg-Val-Asp-Pro-Val-
Miss.All. Ala-Leu-His-Glu-Ala-Val-Asn-His-Ile-Asp-Asp-Ile-Ala-Gly-Ala-Leu-Ser-Arg-Leu-His-Ala-His-Ser-Leu-Arg-Val-Asp-Pro-Val-
Caiman    Ala-Leu-His-Asp-Ala-Val-Asn-His-Ile-Asp-Asp-Ile-Ala-Gly-Ala-Leu-Ser-Arg-Leu-His-Asn-Leu-Arg-Val-Asp-Pro-Val-

                        100                      110                      120
                             π                       π    π   π               π                   π
Human     Asn-Phe-Lys-Leu-Leu-Ser-His-Cys-Leu-Leu-Val-Thr-Leu-Ala-Ala-His-Leu-Pro-Ala-Glu-Phe-Thr-Pro-Ala-Val-His-Ala-Ser-Leu-Asp-Lys-Phe-
Nile Cr.  Asn-Phe-Lys-Phe-Leu-Ala-Gln-Cys-Phe-Leu-Val-Val-Val-Ala-Ile-His-His-Pro-Gly-Ser-Leu-Thr-Pro-Glu-Val-His-Ala-Ser-Leu-Asp-Lys-Phe-
Miss.All. Asn-Phe-Lys-Phe-Leu-Ala-His-Cys-Phe-Leu-Val-Val-Val-Ala-Ile-His-His-Pro-Ser-Ala-Leu-Thr-Pro-Glu-Ile-His-Ala-Ser-Leu-Asp-Lys-Phe-
Caiman    Asn-Phe-Lys-Phe-Leu-Ser-Gln-Cys-Ile-Leu-Val-Val-Phe-Gly-Val-His-His-Pro-Cys-Ser-Leu-Thr-Pro-Glu-Val-His-Ala-Ser-Leu-Asp-Lys-Phe-

                        130                      140   σ    σ
Human     Leu-Ala-Ser-Val-Ser-Thr-Val-Leu-Thr-Ser-Lys-Tyr-Arg
Nile Cr.  Leu-Cys-Ala-Val-Ser-Ser-Val-Leu-Thr-Ser-Lys-Tyr-Arg
Miss.All. Leu-Cys-Ala-Val-Ser-Ala-Val-Leu-Thr-Ser-Lys-Tyr-Arg
Caiman    Leu-Cys-Ala-Val-Ala-Met-Leu-Leu-Thr-Ser-Lys-Tyr-Arg
```

FIG. 1. Amino acid sequence of three crocodilian (Nile crocodile, Mississipi alligator, caiman) hemoglobin α chains. π: α₁β₁ packing contacts; σ: α₁β₂ sliding contacts.

	1	10	20	30

		↓			π
Human	Val-His-Leu-Thr-Pro-Glu-Glu-Lys-Ser-Ala-Val-Thr-Ala-Leu-Trp-Gly-Lys-Val-Asn-Val-Asp-Glu-Val-Gly-Gly-Glu-Ala-Leu-Gly-Arg-Leu-Leu-
Nile Cr.	AcAla-Ser-Phe-Asp-Pro-His-Glu-Lys-Gln-Leu-Ile-Gly-Asp-Leu-Trp-His-Lys-Val-Asp-Val-Ala-His-Cys-Gly-Ala-Glu-Ala-Leu-Ser-Arg-Met-Leu-
Miss. All.	AcAla-Ser-Phe-Asp-Ala-His-Glu-Asp-Lys-Phe-Ile-Val-Asp-Lys-Leu-Trp-Ala-Lys-Val-Asp-Val-Ala-Asp-Cys-Gly-Ala-Glu-Ala-Leu-Ser-Arg-Met-Leu-
Caiman	Ser-Pro-Phe-Ser-Ala-His-Glu-Lys-Glu-Lys-Ile-Val-Asp-Leu-Trp-Val-Asp-Val-Ala-Ser-Cys-Gly-Ala-Ser-Gly-Asp-Ala-Leu-Ser-Arg-Met-Leu-

	40	50	60

	π	π	σ	σ	σ	σ
Human	Val-Val-Tyr-Pro-Trp-Thr-Gln-Arg-Phe-Phe-Glu-Ser-Phe-Gly-Asp-Leu-Ser-Thr-Pro-Asp-Ala-Val-Met-Gly-Asn-Pro-Lys-Val-Lys-Ala-His-Gly-
Nile Cr.	Ile-Val-Tyr-Pro-Trp-Thr-Pro-Arg-Phe-Phe-Ala-Ser-Phe-Gly-Asn-Leu-Ser-Ser-Pro-Thr-Ala-Ile-Met-Gly-Asn-Pro-Lys-Val-Lys-Ala-His-Gly-
Miss. All.	Ile-Val-Tyr-Pro-Trp-Thr-Pro-Arg-Phe-Phe-Gly-His-Phe-Gly-Asn-Leu-Ser-Asn-Ala-His-Ala-Ile-Leu-Asn-Ser-Lys-Val-Gln-Ala-His-Gly-
Caiman	Ile-Ile-Tyr-Pro-Trp-Thr-Pro-Arg-Phe-Phe-Gly-His-Phe-Gly-Asn-Leu-Ser-Ser-Ala-Thr-Ala-Ile-Leu-Asn-Ser-Pro-Lys-Val-Leu-Arg-Glu-His-Gly-

	70	80	90

	σ	σ	σ		↓		↓
Human	Lys-Lys-Val-Leu-Gly-Ala-Phe-Ser-Asp-Gly-Leu-Ala-His-Leu-Asp-Asn-Leu-Lys-Gly-Thr-Phe-Ala-Thr-Leu-Ser-Glu-Leu-His-Cys-Asp-Lys-Leu-
Nile Cr.	Lys-Lys-Val-Leu-Val-Ala-Ser-Phe-Gly-Glu-Ala-Val-Cys-His-Leu-Asp-Asp-Leu-Lys-Gly-Ile-Leu-Arg-Ala-His-Phe-Ala-Asn-Leu-Ser-Lys-Lys-Leu-
Miss. All.	Lys-Lys-Val-Leu-Ala-Ser-Phe-Gly-Glu-Ala-Val-Lys-His-Leu-Asp-Asn-Leu-Lys-Gly-His-Phe-Ala-Ala-Leu-Ser-Lys-Lys-Cys-Lys-Lys-Phe-
Caiman	Lys-Lys-Val-Leu-Ala-Ser-Phe-Gly-Glu-Ala-Val-Lys-His-Leu-Asp-Asn-Ile-Lys-Gly-His-Phe-Ala-His-Leu-Ser-Lys-His-Phe-Glu-Lys-Phe-

	100	110	120

	σ	σ	σ	σ	π	π	π	π
Human	His-Val-Asp-Pro-Glu-Asn-Phe-Arg-Leu-Leu-Gly-Asn-Val-Leu-Val-Cys-Val-Leu-Ala-His-His-Phe-Gly-Lys-Glu-Phe-Thr-Pro-Pro-Val-Gln-Ala-
Nile Cr.	His-Val-Asp-Pro-Glu-Asn-Phe-Lys-Leu-Leu-Gly-Asp-Ile-Ile-Ile-Ile-Val-Leu-Ala-Ala-His-Leu-Pro-Lys-Asp-Phe-Thr-Pro-Lys-Asp-Cys-His-Ala-
Miss. All.	His-Val-Asp-Pro-Glu-Asn-Phe-Lys-Leu-Leu-Gly-Asp-Ile-Ile-Ile-Ile-Val-Leu-Ala-Ala-His-Leu-Pro-Ser-Asp-Phe-Gly-Asp-Ser-His-Ala-
Caiman	His-Val-Asp-Pro-Gly-Asn-Phe-Lys-Leu-Leu-Gly-Asp-Ile-Ile-Ile-Val-Leu-Gly-Met-His-His-Pro-His-Asp-Phe-Thr-Leu-Gln-Thr-His-His-Ala-

	130	140

	π		σ	σ	
Human	Ala-Tyr-Gln-Lys-Val-Val-Ala-Gly-Val-Ala-Asn-Ala-Leu-Ala-His-Lys-Tyr-His
Nile Cr.	Ala-Tyr-Gln-Lys-Leu-Val-Arg-Gln-Val-Ala-Ala-Ala-Leu-Ala-Ala-Glu-Tyr-His
Miss. All.	Ala-Phe-Gln-Lys-Leu-Val-Arg-Gln-Ile-Ala-Ala-Ala-Leu-Ala-Ala-Glu-Tyr-His
Caiman	Ala-Phe-Gln-Lys-Leu-Val-Arg-Val-Ile-Ala-His-Ala-Leu-Ala-Ala-Leu-Ser-Ala-Glu-Tyr-His

FIG. 2. Amino acid sequence of three crocodilian (Nile crocodile, Mississipi alligator, caiman) hemoglobin β chains. σ: $\alpha_1\beta_2$ sliding contacts; π: $\alpha_1\beta_1$ packing contacts; ↓: 2,3-DPG binding site, ↑ HCO_3^- binding site.

Most mammalian hemoglobins are sensitive to 2,3-DPG, that lowers their intrinsic oxygen affinity (Chanutin and Curnish, 1967). This phosphorylated cofactors has been shown to bind between the ends of the β subunits in the deoxyhemoglobin. Its negative charges interact with positive charges from Val 1, His 2 and His 143 of both β chains and from Lys 82 of one β subunit (Arnone, 1972).

Examination of the amino acid sequences of the crocodilian β chains makes it clear that these hemoglobins could not bind the 2,3-DPG. The two His residues are substituted by neutral residues. In Nile crocodile and Mississipi alligator, the N-terminal amino acid is acetylated and thus uncharged. In the caiman β chain, the Pro in position 2 induces a bend in the subunit which takes the charged terminal amino group out of reach of the phosphate group. The only residue left is Lys 82 which is unable to assure the binding alone. For similar reasons carbamino CO_2^- cannot bind to crocodilian hemoglobins. In man, one molecule competes with 2,3-DPG, being in interaction with Val 1β and Lys 82β . A second molecule is fixed to the Val 1α and forms an hydrogen bound with Ser 131α which Ser is substituted by an Ala in the three crocodilian proteins.

It remained to be explained how the two hydrogenocarbonate molecules bind to the crocodilian deoxyhemoglobins. On the basis of the tridimensional model of the human deoxyhemoglobin and of the amino acid replacements encountered in the crocodilian sequences, Perutz investigated all sites able to accomodate stereochemically the HCO_3^- ions, finding only one pair of sites (Perutz et al., 1981). It would appear that the two molecules lodge in the same cavity as the other effectors and are bound to Lys 82 and Glu 144 of one β chain and to the N-terminal residue of the other β subunit. In caiman hemoglobin, the rigid turn brought on by Pro 2 β leads the Ser 1β to form with HCO_3^- a salt bridge through its NH_3^+ and a hydrogen bound through its -OH. In the two others crocodilian hemoglobins a similar binding way occurs provided that a similar turn exists in the polypeptide chain at the level of Ser 2β so that the acetylated α NH- group could form a H- bound with the HCO_3^-. This model should be confirmed by X-ray analysis.

III. AVIAN HEMOGLOBINS

A. Heterogeneity of Avian Hemoglobins

Contrary to crocodilian blood, which shows up only one hemoglobin component, avian blood often displays two hemoglobin components present in unequal proportions, and thus often called the minor and major components. According to the species the minor component accounts for 10-40% of the hemoglobinic material and is often named component 1 as its elutes first from cation exchangers (Vandecasserie et al., 1973) or component D (Brown and Ingram, 1974). For similar reasons the major component is also called component 2 or component A. Some species offer only a single hemoglobin whose properties are nearer those of the major one.

B. Oxygen Affinity

The oxygenation parameters of avian hemoglobins have been examined separately for both components and are somewhat distinct from human hemoglobin. First, Hill coefficient, n, reflecting the cooperativity of the polypeptide chains in the binding of oxygen discloses lower values (n = 2.1 to 2.6) for avian then for mammalian proteins (Vandecasserie et al., 1973). In birds also the intrinsic oxygen affinity of the hemoglobins is lowered by phosphorylated cofactors, usually by inositol pentaphosphate (IP5) (Stewart and Tate, 1969) and in some species, like the ostrich, by both IP5 and inositol tetraphosphate (IP4) (Oberthur et al., 1983a). In the first studies, IP5 was confused with inositol hexaphosphate (IP6) which was still used later on because of its easy availability. It was shown that IP5 is only slightly less effective than IP6 (Brygier and Paul, 1976; Vandecasserie et al., 1976).

The minor components exhibit P_{50} values close to that of the human hemoglobin in the absence as well as in the presence of effector. As for the major or single components, their intrinsic oxygen affinities are higher but the effect of the cofactor is more important, so that in the presence of IP6 their oxygen affinities are lower than that of the human protein (Vandecasserie et al., 1971, 1973).

C. Amino Acid Sequences

The amino acid sequences of five minor and thirteen major avian hemoglobin components have been deciphered, namely chicken (Gallus gallus) (Takei et al., 1975), pheasant (Phasianus colchicus) (Braunitzer and Godovac, 1982), starling (Sturnus vulgaris) (Oberthur and Braunitzer, 1984), ostrich (Struthio camelus) and american rhea (Rhea americana) (Oberthur et al., 1983c) hemoglobin-D and chicken (Knöckel et al., 1982; Matsuda et al., 1971), pheasant (Braunitzer and Godovac, 1982), Canada goose (Branta canadensis) and Mute Swan (Cygnus olor) (Oberthur et al., 1982), greylag goose (Anser anser) (Oberthur et al., 1981), barheaded goose (Anser indicus) (Oberthur et al., 1982), Australian Magpie goose (Anseranas semipalmata) (Oberthur et al., 1983d), Northern Mallard (Anas platyrhynchos platyrhyncos (Godovac and Braunitzer, 1983), American flamingo (Phoenicopterus ruber ruber) (Godovac and Braunitzer, 1984), Starling (Oberthur and Braunitzer, 1984), Golden Eagle (Aquila chrysaetos) (Oberthur et al., 1983b), Ostrich and American rhea (Oberthur et al., 1983a) hemoglobin A.

In all species where the two components have been studied it has been shown that they share the same β chain but differ greatly in the amino acid sequence of their α chain. Figure 3 presents the amino acid sequences of the pheasant hemoglobin α_A and α_D chains and figure 4 that of the β chain. They are aligned to the human homologous subunits. On figures 5, 6, and 7 are only shown the positions where substitutions have been observed among the avian hemoglobin chains. At all other positions, one finds the same residue as in the pheasant hemoglobin.

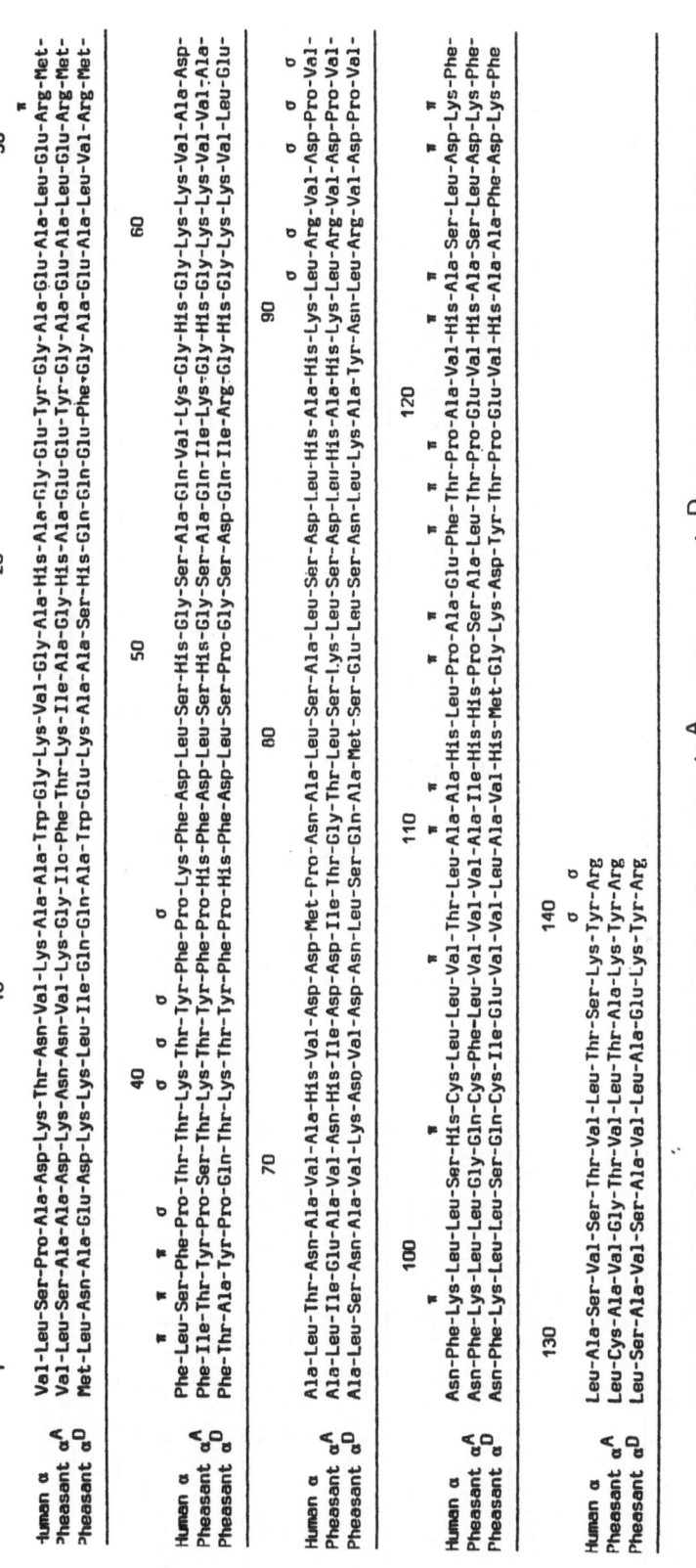

FIG. 3. Amino acid sequence of the α chains of the major (α^A) and the minor (α^D) pheasant hemoglobin components to the human α chain. π: α$_1$β$_1$ packing contacts; σ: α$_1$β$_2$ sliding contacts.

148

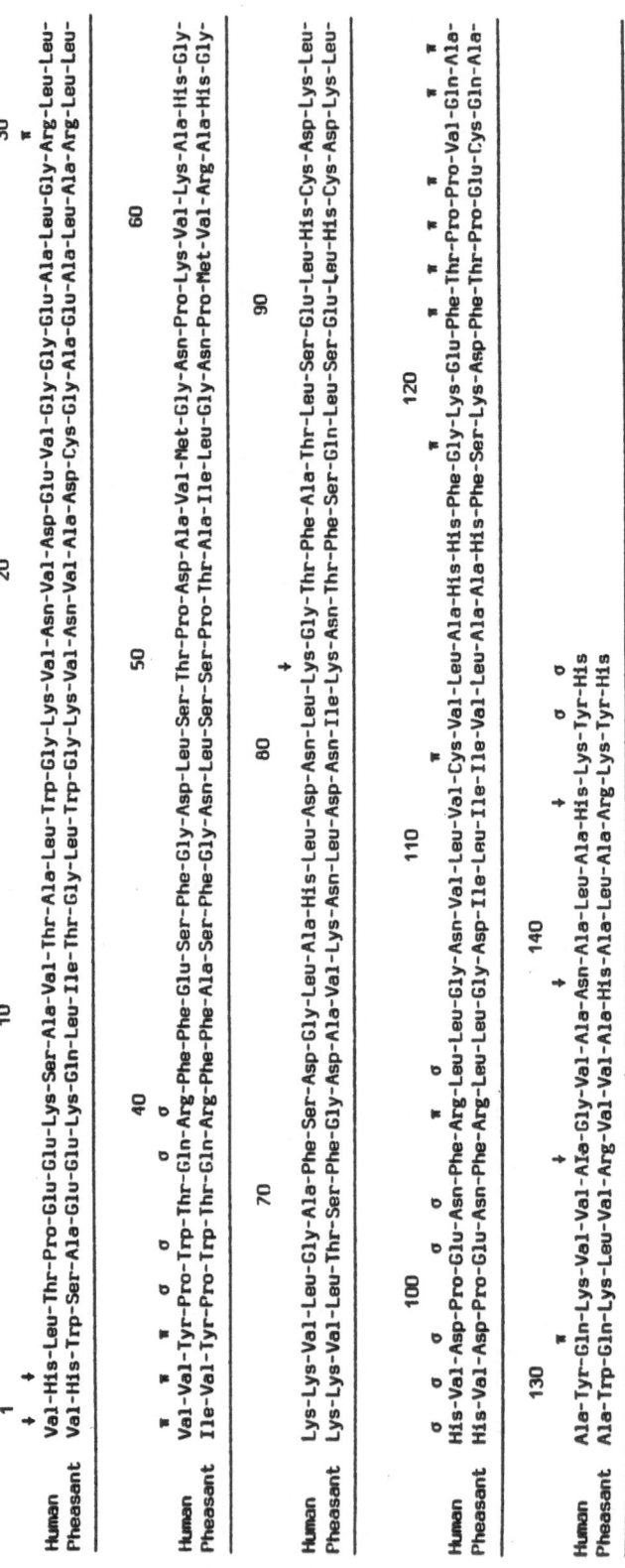

FIG. 4. Amino acid sequence of the unique β chain of the major and the minor pheasant hemoglobin components compared to the human β chain. π: $\alpha_1 \beta_1$ packing contacts; σ: $\alpha_1 \beta_2$ sliding contacts. ↓ IP6 binding site to avian β chain.

FIG. 5. Positions where amino acid substitutions occur in avian αᴬ chains. π: $\alpha_1\beta_1$ packing contacts; σ: $\alpha_1\beta_2$ sliding contacts.

Amino acid position	4	5	8	12	13	15	17	18	19	20	21	22	23	24	25	28	30	34	35	36	38	47	49	50	55
																		π	π	π	σ				
Pheasant	Ala	Ala	Asn	Gly	Ile	Thr	Ile	Ala	Gly	His	Ala	Glu	Glu	Tyr	Gly	Ala	Glu	Ile	Thr	Tyr	Ser	Asp	Ser	His	Ile
Chicken	Ala	Ala	Asn	Gly	Ile	Thr	Ile	Ala	Gly	His	Ala	Gly	Thr	Gln	Thr	Gly	Gln	Ile	Thr	Tyr	Pro	Asp	Ser	His	Ile
Canada goose	Ala	Ala	Thr	Gly	Val	Ser	Ile	Gly	Gly	His	Ala	Asp	Glu	Tyr	Gly	Thr	Glu	Ile	Ala	Ala	Tyr	Asp	Gln	His	Ile
Greylag goose	Ala	Ala	Thr	Gly	Val	Ser	Ile	Gly	Gly	His	Ala	Glu	Glu	Tyr	Gly	Thr	Glu	Ile	Ala	Ala	Tyr	Asp	Gln	His	Ile
Barheaded goose	Ala	Ala	Thr	Gly	Val	Ser	Ile	Ser	Gly	His	Ala	Glu	Glu	Tyr	Gly	Gly	Glu	Ile	Ala	Ala	Tyr	Asp	Gln	His	Ile
Australian magpie goose	Ala	Ala	Gly	Thr	Val	Gly	Ile	Gly	Gly	His	Ala	Glu	Glu	Tyr	Gly	Gly	Gln	Phe	Gln	Gln	Phe	Gln	Gln	Pro	Ile
Northern mallard	Ala	Ala	Thr	Gly	Val	Ser	Ile	Gly	Gly	His	Ala	Glu	Glu	Tyr	Gly	Thr	Glu	Ile	Ala	Ala	Tyr	Asp	Gln	His	Ile
Mute swan	Ala	Ala	Thr	Gly	Val	Ser	Ile	Gly	Gly	His	Ala	Asn	Asp	Tyr	Gly	Thr	Glu	Ile	Ala	Ala	Tyr	Asp	Ser	His	Ile
American flamingo	Ser	His	Ser	Gly	Leu	Gly	Val	Gly	Gly	His	Leu	Glu	Glu	Tyr	Cys	Thr	Ala	Ala	Ala	Ala	Tyr	Gln	Asn	Gln	Pro	Val
Golden eagle	Ala	Asn	Thr	Asn	Val	Thr	Ile	Ser	Gly	Gly	Gln	Ala	Glu	Tyr	Gly	Thr	His	Thr	Pro	His	His	Asn	Gln	Pro	Val
Starling	Ala	Ser	Ala	Ala	Val	Gly	Ile	Gly	Ser	His	Ala	Glu	Glu	Tyr	Gly	Thr	Gln	Ile	Thr	Tyr	Gln	Asp	Gly	Lys	Val
Ostrich	Gly	Thr	Thr	Gly	Ile	Ser	Ser	Ser	His	Ser	Ala	Glu	Glu	Tyr	Gly	Thr	Gln	Ile	Thr	Tyr	Gln	Asp	His	His	Ile
American rhea	Gly	Pro	Thr	Asn	Val	Ala	Ile	Gly	Gly	His	Ala	Asp	Ala	Tyr	Gly	Thr	Gln	Ile	Thr	Tyr	Gln	Asp	His	His	Ile

Amino acid position	57	63	64	67	68	70	72	73	76	77	78	79	88	89	100	103	109	111	113	115	116	119	129	133	134
	π			π		π		π	π				π			π		π	π	π	π	π			
Pheasant	Gly	Val	Ala	Ile	Glu	Val	His	Ile	Thr	Gly	Thr	Ala	His	Leu	Gln	Val	Ile	His	Ile	Ala	Ala	Pro	Leu	Gly	Thr
Chicken	Gly	Val	Ala	Ile	Glu	Val	His	Ile	Ala	Gly	Thr	Ala	His	Leu	Gln	Val	Ile	His	Ala	Ala	Ala	Pro	Leu	Gly	Thr
Canada goose	Ala	Ala	Ala	Val	Glu	Val	His	Ile	Ala	Gly	Ala	Gly	His	Gln	Ala	Val	Ile	Ser	Ala	Ala	Ala	Pro	Leu	Gly	Thr
Greylag goose	Ala	Ala	Ala	Val	Glu	Val	His	Ile	Ala	Gly	Ala	Gly	His	Gln	Ala	Val	Ile	Ser	Ala	Ala	Ala	Pro	Leu	Gly	Thr
Barheaded goose	Ala	Val	Ala	Val	Glu	Val	His	Ile	Ala	Gly	Ala	Gly	His	Gln	Ala	Val	Ile	Ser	Ala	Ala	Ala	Pro	Leu	Gly	Thr
Australian magpie goose	Ala	Ala	Ala	Val	Glu	Val	His	Ile	Ala	Gly	Ala	Gly	His	Gln	Ala	Leu	Ile	His	Ala	Ala	Ala	Met	Gly	Ala	Thr
Northern mallard	Ala	Ala	Ala	Val	Glu	Val	His	Ile	Ala	Gly	Ala	Gly	His	Gln	Ala	Val	Ile	Ser	Ala	Ala	Ala	Pro	Leu	Gly	Ala
Mute swan	Ala	Ala	Ala	Val	Glu	Val	His	Ile	Ala	Gly	Ala	Ser	His	Gln	Ala	Val	Ile	Ser	Ala	Ala	Ala	Pro	Leu	Gly	Ala
American flamingo	Ala	Gly	Ile	Ile	Glu	Val	His	Ile	Met	Gly	Ala	Ser	His	Gln	Val	Ile	Ile	Ser	Ser	Val	His	Ser	Leu	Gly	Thr
Golden eagle	Ala	Val	Gly	Ile	Glu	Val	His	Ile	Ile	Gly	Ala	Gly	His	Gln	Val	Ile	Ile	His	Ser	Val	Asn	Pro	Leu	Gly	Asn
Starling	Gly	Ala	Ala	Asp	Glu	Val	His	Ile	Ser	Gly	Ala	Gly	His	Gln	Leu	Gln	Val	Ile	Ser	Ala	Ser	Ala	Leu	Ser	Ala
Ostrich	Ala	Asn	Asp	Ile	Glu	His	Ile	Ser	Gly	Ala	Ala	Gln	His	Gln	Leu	Ile	Ile	His	Ser	Ala	Ile	Leu	Ser	Gly	Thr
American rhea	Thr	Val	Ser	Ile	Asp	Ala	Asn	Ile	Tyr	Gly	Leu	Ala	Gln	Leu	Gln	Val	Ile	Ser	Ala	Ala	Ile	Leu	Gly	Ala	

Amino acid position	1	3	5	10	11	13	15	17	18	19	20	21	22	23	24	26	30	34	35	38	49	50	53
contact																		π	π	σ			
Pheasant	Met	Asn	Glu	Ile	Gln	Ala	Glu	Ala	Ala	Ser	His	Gln	Gln	Glu	Phe	Ala	Val	Thr	Ala	Gln	Ser	Pro	Asp
Chicken	Met	Thr	Glu	Ile	Gln	Ala	Glu	Ala	Ala	Ser	His	Gln	Glu	Glu	Phe	Ala	Thr	Thr	Gln	Ser		Pro	Asp
Starling	Val	Thr	Glu	Ile	Gln	Thr	Gly	Leu	Gly	Ala	Glu	Glu	Glu	Ile	Ala	Trp	His	Ala	Pro	Ser	Gln		Asp
Ostrich	Met	Thr	Asp	Leu	Gln	Ile	Glu	Val	Gly	Ser	His	Leu	Glu	Asp	Phe	Ala	Glu	Ile	Thr	Gln	His	Pro	Glu
American rhea	Met	Thr	Asp	Ile	Ser	Ile	Thr	Val	Ala	Glu	His	Gly	Gly	Glu	Phe	Gly	Glu	Ile	Thr	Gln	His	Val	Glu

Amino acid position	55	63	64	67	70	72	73	80	81	100	105	108	111	112	113	116	118	123	125	129	130	133	134
contact													π			π		π					
Pheasant	Ile	Leu	Gly	Ser	Val	Asn	Val	Met	Ser	Leu	Ile	Val	His	Met	Asp	Thr	Ala	Phe	Leu	Ser	Ser	Ala	
Chicken	Val	Leu	Gly		Val	Asn	Val	Met	Ala	Leu	Ile	Val	His	Met	Asp	Thr	Ala	Phe	Leu	Ser	Ser	Ala	
Starling	Ile	Val	Ala	Gly	Ile	Ser	Leu	Leu	Ser	Phe	Leu	Thr	Thr	Arg	Leu	Glu	Ser	Val	Met	Ser	Ala	Ala	
Ostrich	Ile	Ala	Asn	Gly	Val	Asn	Leu	Met	Ser	Leu	Phe	Val	His	Met	Asp	Thr	Ala	Tyr	Leu	Thr	Ala	Ala	
American rhea	Val	Val	Asn	Ser	Val	Asn	Leu	Ala	Leu	Phe	Val	His	Leu	Leu	His	Leu	Thr	Ala	Tyr	Leu	Ser	Ala	Ser

FIG. 6. Positions where amino acid substitutions occur in avian α^D chains. π: $\alpha_1\beta_1$ packing contacts; σ: $\alpha_1\beta_2$ sliding contacts.

Amino acid position: 2 4 12 13 22 43 54 55 59 61 73 80 86 111 112 119 121 123 125 136

(markers: ↓ at 2; σ at 43; π at 55; π at 112; π at 119; π at 123; π at 125)

Pheasant	His¹-Ser-Thr-Gly-Asp-Ala-Ile-Leu-Met-Arg-Asp-Asn-Ser-Ile-Ser-Asp-Thr-Glu-Val-
Chicken	His-Thr-Thr-Gly-Glu-Ala-Ile-Leu-Met-Arg-Asp-Asn-Ser-Ile-Ser-Asp-Thr-Glu-Val-
Canada goose	His-Thr-Thr-Gly-Asp-Ser-Ile-Leu-Met-Arg-Asp-Asn-Ala-Ile-Ala-Asp-Thr-Asp-Val-
Greylag goose	His-Ser-Thr-Gly-Asp-Ser-Ile-Leu-Met-Arg-Asp-Asn-Ala-Ile-Glu-Thr-Glu-Val-
Bar-headed goose	His-Ser-Thr-Gly-Asp-Ser-Ile-Leu-Met-Arg-Asp-Asn-Ala-Ile-Ala-Glu-Thr-Asp-Val-
Australian magpie goose	His-Ser-Thr-Gly-Asp-Ser-Ile-Leu-Met-Arg-Asp-Asn-Ala-Ile-Ala-Glu-Thr-Asp-Val-
Northern mallard	His-Thr-Thr-Gly-Asp-Ala-Ile-Leu-Met-Arg-Asp-Asn-Ala-Ile-Ile-Pro-Glu-Thr-Asp-Val-
Mute swan	His-Thr-Thr-Gly-Asp-Ser-Ile-Leu-Met-Arg-Asp-Asn-Ala-Ile-Ala-Asp-Thr-Asp-Val-
American flamingo	His-Ser-Thr-Ser-Asp-Ala-Ile-Leu-Met-Arg-Asp-Asn-Ala-Ile-Ile-Ala-Asp-Thr-Asp-Ala-
Golden eagle	His-Thr-Thr-Gly-Asp-Ala-Ile-Leu-Met-Arg-Glu-Asn-Ala-Ile-Ile-Thr-Asp-Ser-Asp-Ala-
Starling	Gln-Thr-Thr-Gly-Gln-Ala-Val-Leu-Lys-Gln-Asp-Ser-Ser-Val-Gly-Asn-Ala-Ile-Thr-Glu-Thr-Asp-Val-
Ostrich	Gln-Ser-Ser-Gly-Asp-Ala-Ile-Leu-Met-Arg-Asp-Asn-Ala-Ile-Ile-Thr-Glu-Thr-Glu-Val-
American rhea	Gln-Thr-Thr-Gly-Asp-Ala-Ile-Leu-Met-Arg-Asp-Asn-Ala-Ile-Ile-Ala-Asp-Thr-Glu-Val-

FIG. 7. Positions where amino acid substitutions occur in avian β chains. π: $\alpha_1\beta_1$ packing contacts; σ: $\alpha_1\beta_2$ sliding contacts; ↓ IP6 binding site.

D. Relation between Structure and Activity

The residues at the IP6 binding site have been identified by Arnone and Perutz (1974) on the human hemoglobin. They are the same as those already involved in the binding of 2,3-DPG. In birds hemoglobins position 143 is occupied by Arg rather than His. Two others basic residues appear in the internal cavity, Arg instead of Ala 135β and His instead of Asn 139β and could contribute to neutralizing the phosphate charges and so indirectly increase the binding energy. Indeed the entropy variation that goes with the binding of IP6 to pigeon hemoglobin is much higher (-21,9 kcal/mole) (Vandecasserie et al., 1974) than that of the binding of 2,3-DPG to human hemoglobin (-13,2 kcal/mole) (Benesch et al., 1967).

In a few species His 2β is replaced by Gln, but this residue seems to be able to bind the cofactor just as well (Oberthur et al., 1983a).

Besides its function in diminishing hemoglobin oxygen affinity, a quite new role has recently been proposed for the phosphorylated cofactor in avian blood (Baumann et al., 1984). It is well known that hemoglobin solubility in the red blood cells must be very high. Actually, in the deoxyform the isolated avian hemoglobin D, as well as the chicken embryonic hemoglobin, crystallize well before their physiological concentrations. However those components solubilize by the addition of either organic phosphates or avian major hemoglobinic components. In embryonic red blood cells solubilization must rely entirely on phosphorylated cofactor, because the hemoglobin A concentration is too low; in adult erythrocytes, solubilization could already be achieved by hemoglobin A and could be due to a joined effect of hemoglobin A and organic phosphate. The low solubility of hemoglobin D should rely on some particular details of their α chains as they shared with hemoglobin A the same β chain. All known α^D sequences present residues in the sliding contact $\alpha_1 \beta_2$ residues identical or very similar to those of avian α^A or human α^A (Fig. 8). Nevertheless, close to the C-terminal side of the chains, in position 138, one finds a neutral residue, i.e. Ser in most mammalian, Ala in all avian α^A chains but a negatively charged Glu in all α^D and in chicken embryonic π chain (Engel et al., 1983). The authors suggest that this substitution could elicit a site for intermolecular interactions.

A nice example of molecular adaptation was described by the joint studies of Petschow et al. (1977) and Rollema and Bauer (1979). It concerned the relation between the hemoglobin oxygen affinities and molecular structural of two goose species, the greylag goose and the bar-headed goose. The first lives in plains and has a normal oxygen affinity. The second crosses the Himalayas and presents a higher oxygen affinity. One single substitution from the usually invariant Pro 119 α, important in the $\alpha_1 \beta_1$ contact (Fermi and Perutz, 1981) to Ala in bar-headed goose (Oberthur et al., 1981) results in a very slight increase in its oxygen affinity. This increase is amplified 10-fold by the addition of IP5, explaining the high oxygen affinity of its blood.

Amino acid position	37(T)	38	40(T)	41	42	44(T)	91(T)	92	94	95	96	97	138	140	141
Human	Pro	Thr	Lys	Thr	Tyr	Pro	Leu	Arg	Asp	Pro	Val	Asn	Ser	Tyr	Arg
Avian α^A	Pro	Pro / Ser / Gln	Lys	Thr	Tyr	Pro	Leu	Arg	Asp	Pro	Val	Asn	Ala	Tyr	Arg
Avian α^D	Pro	Gln / Pro	Lys	Thr	Tyr	Pro	Leu	Arg	Asp	Pro	Val	Asn	Glu	Tyr	Arg
Avian embryonic π	Pro	Gln	Lys	Thr	Tyr	Pro	Leu	Arg	Asp	Pro	Val	Asn	Glu	Tyr	Arg
Avian β	Pro	Tyr	Gln	Arg	Ala / Ser	His	Val	Asp	Glu	Asn	Leu	Tyr	His		
Human β	Pro	Tyr	Gln	Arg	Glu	His	Val	Asp	Glu	Asn	Leu	Tyr	His		
Amino acid position	36	37	39(R)	40	43	97	98(T)	99	101	102(R)	105(T)	145(T)	146(T)		

FIG. 8. Amino acid residues involved in 138α sliding contacts. Near these residues in position substitution of a neutral into a negatively charges side chain may induce a rearrangement of the C-terminus and explain the low solubility of embryonic and minor avian hemoglobin components.

153

IV. MOLECULAR EVOLUTION

We have just seen that among the numerous amino acid substitutions that appear between hemoglobins of various species very few seem functionally important and are necessary to explain a molecular adaptation. So, most of the amino acid replacements appear to be selectively neutral, caused by random drift (Kimura, 1979).

A simple count of amino acid substitutions or nucleotide mutations is often used to build a phylogenetic tree based on those molecular data. However, it seems that the two kinds of polypeptide chains did not evolve at the same rate. For instance if one compares the number of amino acid substitutions between the α or the β chains among the various crocodilian hemoglobins, and the number of substitutions of the crocodilian chains to those of the homologous human polypeptides (Table 1), one notices that the α chains evolved more slowly than the β chains. On the other hand, if one counts the number of amino acid replacements between the avian α_A, α_D or β chains (Tables 2 and 3) the reverse situation is observed, the β chains having evolved more slowly than the α chains. So it seems that some external factors may either accelerate or slow down the mutation rate. One could imagine that the avian β hemoglobin chains have a lower degree of freedom because they have to accommodate two types of α chains.

TABLE 1. Number of aminoacid substitutions between α (above the diagonal) and β (below the diagonal) crocodilian and human hemoglobin chains.

	Human	Nile crocodile	Mississipi alligator	Caiman	
Human		44	47	46	
Nile crocodile	66		17	21	α chains
Mississipi alligator	75	28		20	
Caiman	74	40	29		

β chains

Moreover it is possible that the mutation rates are not always the same along the different branches of the phylogenetic tree. One finds fewer amino acid substitutions between avian and crocodilian hemoglobins than between mammalian and crocodilian hemoglobins, but even more differences between viper and crocodilian α chains (Perutz et al., 1981). Is there any mistake in the taxonomy or is it more likely as it is also suggested by Romero-Errera et al. (1978) that proteins do not always evolve at the same rate, but depend on external contraints ?

Some more data will be needed to refine these approaches.

TABLE 2. Number of amino acid substitutions between α^A (above the diagonal) and β (below the diagonal) avian hemoglobin chains

Avian α^A chains / Avian β chains

	Chicken	Pheasant	Canada goose	Greylag goose	Barheaded goose	Australian magpie goose	Northern mallard	Mute swan	Golden eagle	American flamingo	Starling	Ostrich	American rhea
Chicken		6	19	17	17	21	18	21	18	28	22	20	22
Pheasant	2		19	18	18	25	17	20	18	31	24	18	24
Canada goose	5	5		2	5	14	6	3	19	23	21	16	20
Greylag goose	6	4	2		3	12	5	4	17	22	20	15	20
Barheaded goose	7	5	5	1		3	8	7	16	25	23	16	21
Australian magpie goose	6	4	14	4	3		16	15	24	22	20	23	24
Northern mallard	4	4	6	3	4	6		5	20	26	21	15	22
Mute swan	5	5	3	3	2	3	5		18	25	23	15	19
Golden eagle	8	8	6	9	8	6	7	6		31	26	17	17
American flamingo	6	4	5	4	5	4	5	5	7		27	29	31
Starling	9	11	11	14	13	13	12	11	14	14		26	25
Ostrich	7	5	7	4	5	7	4	7	9	6	13		18
American rhea	4	4	3	4	5	6	3	3	7	4	10	4	

TABLE 3. Number of amino acid substitutions between avian α^D hemoglobin chains

	Chicken	Pheasant	Starling	Ostrich	American rhea
Chicken					
Pheasant	7				
Starling	34	34			
Ostrich	20	22	34		
American rhea	24	28	36	19	

REFERENCES

Arnone A (1972) X-ray diffraction study of binding 2,3-diphosphoglycerate to human deoxyhaemoglobin. Nature 237: 146-149

Arnone A, Perutz M (1974) Structure of inositol hexaphosphate - human deoxyhaemoglobin complex. Nature 249: 34-36

Bauer C, Forster M, Gros C, Mosca A, Perella M, Rollema HS, Vogel D (1981) Analysis of bicarbonate binding to crocodilian hemoglobin. J. Biol. Chem. 256: 8429-8435

Bauer C, Jelkmann W (1977) Carbon dioxide governs the oxygen affinity of crocodile blood. Nature 269: 825-827

Baumann R, Goldbach E, Haller EA, Wright PG (1984) Organic phosphates increase the solubility of avian haemoglobin D and embryonic chicken haemoglobin. Biochem. J. 217: 767-771

Benesch R, Benesch RE (1967) The effect of organic phosphates from the human erythrocyte on the allosteric properties of hemoglobin. Biochem. Biophys. Res. Commun. 26: 162-167

Braunitzer G, Godovac J (1982) The amino acid sequence of pheasant (Phasianus colchicus colchicus) hemoglobins. Hoppe-Seyler's Z. Physiol. Chem. 363: 229-238

Brown JL, Ingram VM (1974) Structural studies on chick embryonic hemoglobins. J. Biol. Chem. 249: 3960-3972

Brygier J, Paul C (1976) Oxygen equilibrium of chicken hemoglobin in the presence of organic phosphates. Biochimie 58: 755-756

Chanutin A, Curnish RR (1967) Effect of the organic and inorganic phosphates on the oxygen equilibrium of human erythrocytes. Arch. Biochem. Biophys. 121: 96-102

Dill DB, Edwards HT (1931) Oxygen affinity of crocodilian blood. J. Biol. Chem. 90: 5115-5130

Engel JD, Rusling DJ, McCurne KC, Dodgson JB (1983) Unusual structure of the chicken embryonic α globin gene π'. Proc. Natl. Acad. Sci. USA 80: 1392-1396

Fermi G (1975) Three-dimensional Fourier synthesis of human deoxyhaemoglobin at 2.5 Å resolution: refinement of the atomic model. J. Mol. Biol. 97: 237-256

Fermi G, Perutz MF (1981) Haemoglobin and Myoglobin. Atlas of Biological Structures. Vol. 2, Clarendon, Oxford.

Godovac-Zimmermann J, Braunitzer G (1983) The amino acid sequence of northern mallard (Anas platyrhynchos platyrhynchos) hemoglobin. Hoppe-Seyler's Z. Physiol. Chem. 364: 665-674

Godovac-Zimmermann J, Braunitzer G (1984) The amino acid sequence of α^A and β chains from the major hemoglobin component of american flamingo (Phoenicopterus ruber ruber). Hoppe-Seler's Z. Physiol. Chem. 365: 437-443

Jelkmann W, Bauer C (1980) Oxygen binding properties of caiman blood in the absence and presence of carbon dioxide. Comp. Biochem. Physiol. 65A: 331-336.

Kimura M (1979) The neutral theory of molecular evolution. Sci. Amer. 241: 94-104

Knöckel N, Wittig B, Wittig S, John ME, Grundmann U, Oberthür W, Godovac J, Braunitzer G (1982) No evidence for "stress" α-globin genes in chicken. Nature 295: 710-712

Leclercq F, Bauer C, Fraboni A, Paul C, Vandecasserie C, Schnek AG, Braunitzer G (1980) Caiman crocodylus Hemoglobin. Structure and Activity In: Peeters H (ed) Protides and Biological Fluids. 28th Colloquium. Pergamon Press, New York, pp. 79-82

Leclercq F, Schnek AG, Braunitzer G, Stangl A, Schrank B (1981) Direct reciprocal allosteric interaction of oxygen and hydrogen carbonate. Sequence of the haemoglobins of the caiman (Caiman crocodylus), the Nile crocodile (Crocodylus niloticus) and the Mississipi crocodile (Alligator mississipiensis). Hoppe-Seyler's Z. Physiol. Chem. 362: 1151-1158

Matsuda G, Maita T, Mizuno K, Ota H (1973) Amino-acid sequence of A_{II} component of adult chicken haemoglobin. Nature New Biology 244: 244

Matsuda G, Takei H, Wu KC, Shiozawa T (1971) The primary structure of the α polypeptide chain of A_{II} component of adult chicken hemoglobin. Int. J. Prot. Res. 3: 173-174

Oberthür W, Braunitzer G (1984) Hämoglobin vom gemeinen star (Sturnus vulgaris, passeriformes): Die primarstrukturen der α^A und β ketten. Hoppe-Seyler's Z. Physiol. Chem. 365: 159-173

Oberthür W, Braunitzer G, Kalas S (1981) Untersuchungen am hämoglobin der graugans (Anser anser). Die primarstruktur der α- und β-ketten der hauptkomponents. Hoppe-Seyler's Z. Physiol. Chem. 362: 1101-1112

Oberthür W, Braunitzer G, Würdinger I (1982) Das hämoglobin der streifengans (Anser indicus) Primärstruktur und physiologie der atmung, systematik und evolution. Hoppe-Seyler's Z. Physiol. Chem. 363: 581-590

Oberthür W, Braunitzer G, Baumann R, Wright P (1983a) Die primärstruktur der α- und β-ketten der hauptkomponenten der hämoglobine der straubes (Struthio camelus) und des nandus (Rhea americana) (Struthio formes). Aspekte zur almungs physiologie und systematik. Hoppe-Seyler's Z. Physiol. Chem. 364: 119-134

Oberthür W, Braunitzer G, Grimm F, Kösters J (1983b) Hämoglobine des steinadlers (Aquila chrysaetos, accipitriformes): Die aminosaure-sequenz der α^A und β^B-ketten der hauptkomponente. Hoppe-Seyler's Z. Physiol. Chem. 364: 851-858

Oberthür W, Godovac-Zimmermann J, Braunitzer G (1982) The amino-acid sequence of canada goose (Branta canadensis) and mute swan (Cygnus olor) Hemoglobins. Two different species with identical β chains. Hoppe-Seyler's Z. Physiol. Chem. 363: 777-787

Oberthür W, Godovac-Zimmermann J, Braunitzer G (1983c) The different evolution of bird hemoglobin chains in hemoglobin. Brussels Hemoglobin Symposium. Schnek AG, Paul C (eds) Editions de l'Université de Bruxelles, pp. 365-375

Oberthür W, Wiesner H, Braunitzer G (1983d) Die primarstruktur der α- und β-ketten der hauptkomponente der hämoglobine der spattfubgans (<u>Anseranas semipalmata</u>, anatidae). Hoppe-Seyler's Z. Physiol. Chem. 364: 51-59

Perutz MF (1983) Species adaptation in a protein molecule. Mol. Biol. Evol. 1: 1-28

Perutz MF, Bauer C, Gros G, Leclercq F, Vandecasserie C, Schnek AG, Braunitzer G, Friday AE, Joysey KA (1981) Allosteric regulation of crocodilian hemoglobin. Nature 291: 682-684

Petschow D, Wurdinger I, Baumann R, Duhm J, Braunitzer G, Bauer C (1977) Causes of high blood O_2 affinity of animals living at high altitude. J. Appl. Physiol. 42: 139-143

Rapoport S, Guest GM (1941) Distribution of acid soluble phosphorus in the blood cells of various vertebrates. J. Biol. Chem. 138: 269-282

Rollema HS, Bauer C (1979) The interaction of inositol pentaphosphate with the hemoglobins of highland and lowland geese. J. Biol. Chem. 254: 12038-12043

Romer AS (1966) Vertebrate paleontology. Chicago-London: University of Chicago Press

Romero-Herrera AE, Lehmann H, Joysey KA, Friday AE (1978) The use of amino-acid sequence analysis in assessing evolution. Phil. Trans. R. Soc. B. 283: 61-163

Steward JH, Tate ME (1969) Gel chromatography of inositol polyphosphates and the avian haemoglobin-inositol pentaphosphate complex. J. Chromatogr. 45: 400-406

Takei H, Ota Y, Wu K, Kiyohara I, Matsuda G (1975) Amino acid sequence of the α chain of chicken AI Hemoglobin. J. Biochem. 77: 1345-1347

Vandecasserie C, Fraboni A, Schnek AG, Léonis J (1976) Oxygen affinity of some avian hemoglobins in presence of various phosphorylated cofactors. Colloque sur l'Hémoglobine (Le Touquet - Paris - Plage; 25 mai 1976) p. 34

Vandecasserie C, Paredes S, Schnek AG, Léonis J (1974) Etude calorimétrique de la fixation de l'inositolhexaphosphate sur l'hémoglobine de pigeon. Arch. Intern. Physiol. Bioch. 82: 1021-1023

Vandecasserie C, Paul C, Schnek AG, Léonis J (1973) Oxygen affinity of avian hemoglobins. Comp. Biochem. Physiol. 44A: 711-718

Vandecasserie C, Schnek AG, Léonis J (1971) Oxygen affinity studies of avian hemoglobins. Chicken and pigeon. Eur. J. Biochem. 24: 284-287

Abnormal Human Hemoglobins:
Molecular Tools to Study Normal Hemoglobin Functions

C. POYART, E. BURSAUX, F. GALACTEROS, H. WACJMAN

I. INTRODUCTION

Starting with a basic ancestral structure natural selection has provided various ways by which different living species have adapted hemoglobin functions to their respiratory requirements, oxygen transport and the buffering of protons released in the reaction of carbon dioxide with water.

In most mammalian species these adaptations are accounted for by punctual changes of key residues in the primary structure of the globin chains and also through the binding of heterotropic cofactors, chloride and organic phosphates (Perutz and Imaï, 1980; Perutz and Brunori, 1982; Perutz, 1983). The presence of these cofactors in the red cells leads to a decrease of the oxygen affinity of the native hemoglobin (by stabilizing the T deoxy quaternary structure) and to an increased in the alkaline Bohr effect by increasing the pK of the residues to which they bind (Benesch and Benesch, 1974). The effects of anions on hemoglobin are oxygen linked due to their differential binding between the fully oxy and fully deoxy configurations.

Genetic Hb variants which result from a point substitution of a natural residue in the globin chains have been, in the past 10 years, of great interest to understand and/or to confirm the detailed and fine tuning of normal Hb functions as postulated from the stereochemical theory of Perutz (1968, 1970). Some of these mutant Hbs of which approx. 550 have been discovered (IHIC, 1984), have been studied carefully in their abnormal functions and in their 3-D structures obtained from X-ray crystallography. This has helped to recognize or to confirm the critical role of specific domains within the tetramer and of key residues involved in regulating Hb function (Kilmartin et al., 1978, Perutz et al., 1980).

1) Residues lining the heme pocket and involved in heme contacts where substitutions lead to heme instability, methemoglobin formation and anaemia (Bonaventura et al., 1978; Bunn et al., 1977; Fermi, 1981).

2) Residues at the N-C termini of the α and β chains responsible for the major part of the alkaline Bohr effect (Kilmartin and Wootton, 1970; Kilmartin et al., 1977; Kilmartin et al., 1978; Kilmartin et al., 1980).

3) Residues situated at the $\alpha_1\beta_2$ interface between the two $\alpha\beta$ dimers where normal structure is necessary for the correct transition between the T and R quaternary configurations responsible for normal cooperativity in ligand binding (Perutz and Lehmann, 1968; Pettigrew et al., 1982).

4) Residues lining the $\alpha_1\beta_2$ cleft which are the sites for organic phosphate binding and partly responsible for the alkaline and acid Bohr effect (Arnone, 1972; Perutz et al., 1980).

5) And as a whole the importance of internal residues in electrostatic and hydrophobic interactions compared to external mostly polar residues where mutations rarely affect Hb function. The important exception to this statement is sickle cell Hb which is the most frequent Hb variant encountered in the human population. In this Hb the external β_6 Glu is substituted for a valyl residue. This does not influence Hb function when diluted in vitro. At the high concentration inside the red cells the mutation leads to polymerisation of the Hb tetramers when the pO_2 is decreased below a certain level. This is responsible for gross alteration of the red cells viscosity and mechanical properties and severe clinical disease.

We shall give in the following few examples of mutations at some specific sites cited above.

II. MUTATIONS AT THE $\alpha_1\alpha_2$ N-C TERMINAL INTERFACE

The first example concerns the physiological importance of the N-C terminal region of the $\alpha_1\alpha_2$ interface (Fig. 1). X-rays crystallographic data as well as biochemical studies (Arnone, 1977; O'Donnell et al., 1979; Perutz, 1970) have revealed the complex network of salt bridges and hydrogen bonds linking the N-Val α_1 to the side chain of C-Arg 141 α_2 and other neighboring residues. Arnone (1977) has shown that a small ion like chloride makes the salt bridge between the amino group of Val α_1 and the guanidinium group of Arg 141 α_2. Biochemical modifications at the N-terminus (Kilmartin et al., 1977; O'Donnell et al., 1979) or cutting off the arginyl residue through CPB digestion leads to an increased oxygen affinity, decreased cooperativity, decreased alkaline Bohr effect and low chloride binding. Table 1 shows the 5 Hb variants which have been reported up to now with substitutions at the arginine 141α (IHIC, 1984). At least two of these variants, Hb Suresnes (Poyart et al., 1980) and Hb Legnano for which structural and functional studies have been reported confirm the abnormalities due to the suppression of Arg 141 α. This is shown in

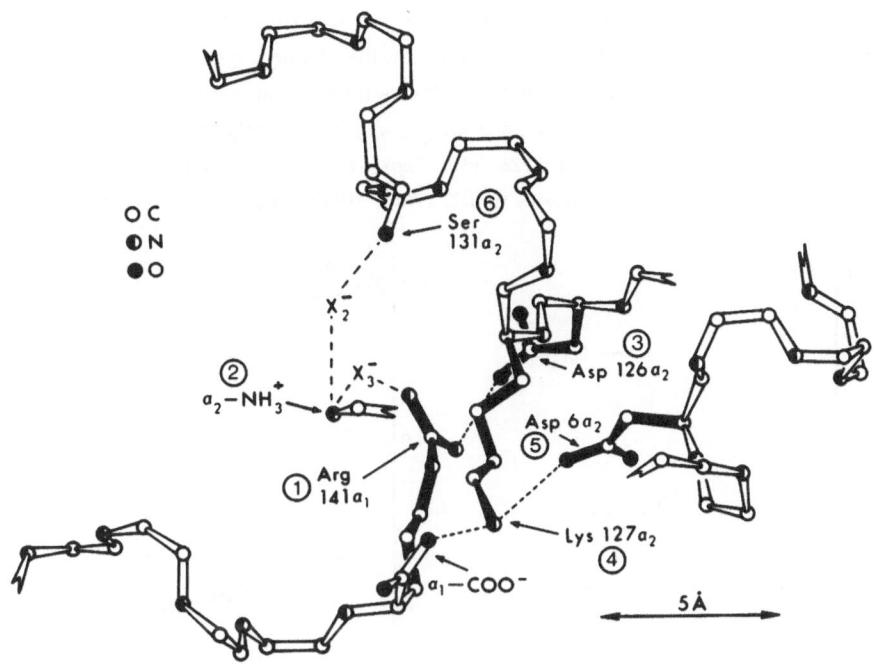

FIG. 1. Sketch of the $\alpha_1 \alpha_2$ contacts showing the salt bridge between α_1 1 (NA$_2$) Val amino-group and α_2 141 (HC$_3$) Arg through anion binding (from Arnone, 1977).

TABLE 1. Hemoglobin Variants at the $\alpha_1 \alpha_2$ Chains Interface

	141 (HC$_3$) Arg $\longrightarrow$
Hb Singapore	Pro
Hb Suresnes	Arg
Hb Legnano	Leu
Hb Cubujuqui	Ser
Hb Camagüey	Gly
Hb Tarrant	126 (H$_9$) Asp $\longrightarrow$ Asn

162

figures 2 and 3 which illustrate the abnormal function of Hb Suresnes. In all the experimental conditions studied this mutant retains higher oxygen affinity, lower alkaline Bohr effect and about half oxygen linked chloride binding compared to HbA. In addition the cooperativity in the absence of anions was low and partially restored upon addition of organic phosphate. Hb Tarrant (α126 (H$_9$) Asp$\longrightarrow$Asn, site 2 in figure 1) has also a high oxygen affinity and low cooperativity (IHIC, 1984). This residue is also a partner of an

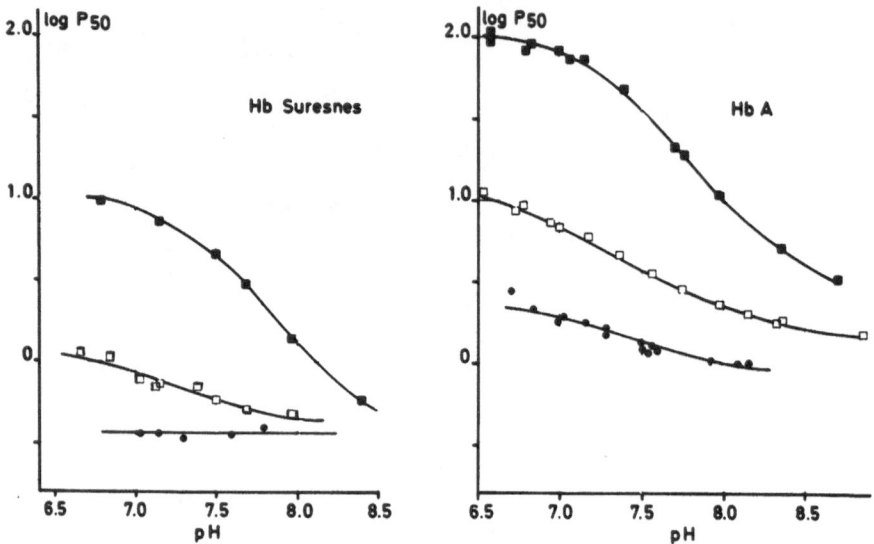

FIG. 2.   Changes in oxygen affinity (Log P$_{50}$, mmHg) with pH in Hb Suresnes and HbAo. Filled circles: 5 mM chloride; open squares: 100 mM chloride; filled squares: 1.5 mM Inositol hexaphosphate.
Conditions: 0.05 M bis-Tris or Tris buffer, 0.6 mM Heme, T 25 °C (Poyart et al., 1980).

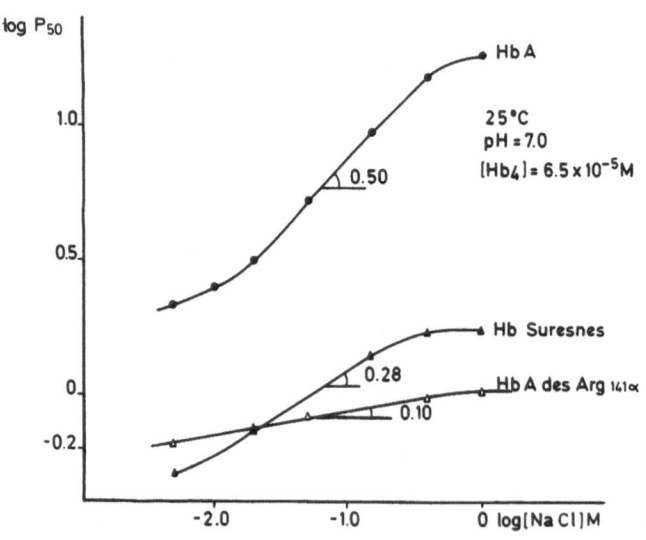

FIG. 3.   Titration with chloride of purified HbAo, Hb Suresnes and CPB digested HbAo (des Arg 141 α) (Poyart et al., 1980).

important $\alpha_1 \beta_1$ contact through H bonds with β_{34}Val and β_{35}Tyr. By contrast mutations at sites 4 and 5 do not lead to hematological disorders. No mutations have been so far described at site 6 (α131 (H$_{14}$) Ser) nor at α_1(NA$_2$) Val where only elongations of the α chains have been described (IHIC, 1984).

These results point to the important role of Arg 141 α and of the α_1-α_2 salt bridge through chloride binding in stabilizing the T structure of the deoxy tetramer and for the full expression of the alkaline Bohr effect. The abnormality brought about by the substitution of Arg 141 α remains localized to this specific domain without altering irreversibly the other functional properties of the molecule. This is borne out by the partial restoration of cooperativity by organic phosphate as well as that of the alkaline Bohr effect (Poyart et al., 1980).

III. MUTATIONS AT THE C-TERMINAL REGION OF THE β CHAINS

In HbA, the C-terminal β 146 (HC$_3$) residue is a histidine of which the imino NH$^+$ is in oxygen linked electrostatic interaction with the neighboring γ carboxylate of β 94 (FG$_1$) Asp of the same chain (Perutz, 1970). The carboxyl group of the same histidine makes also a salt bridge with the amino group of the α_{40} (C$_5$) Lysine, in the deoxy configuration (Fig. 4). These interactions are ruptured upon oxygenation and account, at neutral pH, for 40 to 60 per cent of the alkaline Bohr effect (Kilmartin et al., 1978; Perutz et al., 1980). Several Hb variants have been described at the β_{146} (HC$_3$) His residue (1 in figure 4) which all show increased oxygen affinity and lowered Bohr effect (Table 2) (IHIC, 1984). A similar picture was observed in mutant Hb at the β94 (FG$_1$) Asp residue, and in Hb Kariya (α_{40} (C$_5$) Lys $\longrightarrow$ Glu), the salt bridge partners of β 146 (HC$_3$) His (sites 2 and 4 in figure 4). More recently this was also confirmed by the discovery of Hb Okasaki (Table 2) where the β 93 (F$_9$) cystyl residue is replaced by an arginyl residue (Harano et al., 1984). In this case the decreased Bohr effect observed may be accounted for the weakening or the disruption of the β (HC$_3$) His - β(FG$_1$) Asp salt bridge by the bulky arginyl side chain. A similar alteration was reported upon binding N-ethylmaleimide compound to the -SH group of the β93 (F$_9$) Cys. All these studies have clearly established that the C-terminal salt bridge in the β chain is at the origin of about half the alkaline Bohr effect.

An other important aspect in the functional properties of these mutant Hb is, as indicated in Table 2, that anion binding either chloride or DPG, is preserved inspite of widely varying cooperativities and oxygen affinities which illustrates the localized character of the functional abnormality due to the alteration of the C-terminal intra β chain salt bridge.

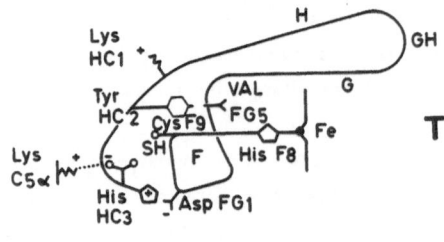

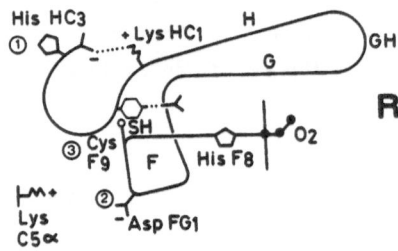

FIG. 4. Schematic representation of the C-terminal β intra chain salt-bridges, in the deoxy T (upper) and in the oxy R (lower) configurations (from Perutz, 1983).

TABLE 2. Hemoglobin Variants at the C Terminus Intrachain β Salt Bridge

	β 146 (HC3) His →	Oxygen affinity	n_{50}	Bohr effect	Anion binding
Hb Hiroshima	Asp	↗	↘	↘	± ↘
Hb York	Pro	↗	↘	↘	Nl
Hb Cochin-Port Royal	Arg	↗±	Nl	↘	Nl
Hb Cow town	Leu	↗	Nl	↘	±Nl
$\alpha_2 \beta_2$Des-His 146		↗	↘	↘	
Hb Barcelona	β 94 (FG$_1$)ASP → His	↗	↘	↘	Nl
Hb Bunbury (unfractionated)	β 94 (FG$_1$) Asp → Asn	↗	↘	↘	?
Hb Okazaki	β 93 (F$_9$) Cyst → Arg	↗	↘	↘	±Nl
Hb Kariya	α 40 (C$_5$) Lys → Glu	↗	↘	↘	↘

Nl = normal ⎫
                 ⎬ compared to HbAo
± = slightly ⎭

IV. MUTATIONS AT THE $\beta_1 \beta_2$ GLOBIN CHAINS INTERFACE

Figure 5 shows the structure of the entrance of the inter β chains cavity, lined with a cluster of positively charged residues (Val 1 β, His 2 β, His 143 β and Lys 82 β). This domain in the molecule was demonstrated by Arnone (1972) as the site for the organic phosphate binding in the deoxy conformation. Upon binding to the deoxy HbA, 2,3 DPG decreases the oxygen affinity of the native molecule a factor of 3 to 4 fold at neutral pH and increases the alkaline Bohr effect (Benesch and Benesch, 1974). Mutant Hb at one of these binding sites have all demonstrated that the various substitutions lead to an inhibition of the effect of DPG and varying degree of inhibition of the alkaline Bohr effect compared to HbA. It has been also demonstrated that this part of the molecule should be a binding site for chloride as

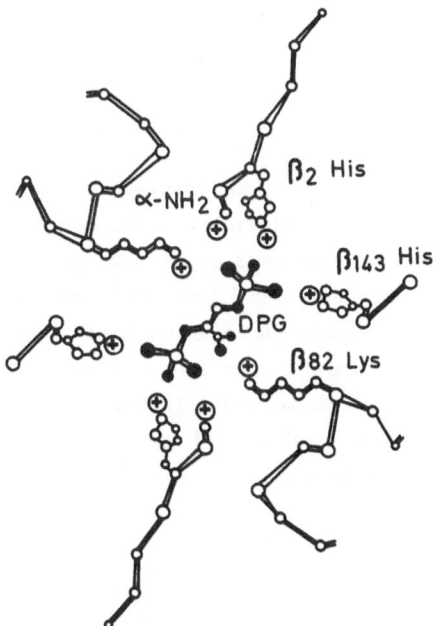

FIG. 5. Schematic drawing of the diphosphoglycerate binding showing the cluster of positively charged residues lining the entrance of the central cavity between the two β chains in deoxy HbAo (from Arnone, 1972).

indicated by the competition between chloride and organic phosphates illustrated in figure 6 (Benesch and Benesch, 1974; Bonaventura and Bonaventura, 1978; Manning and Nigen, 1978). As discussed by Perutz et al. (1980) mutant Hb have been of great interest to assign among these 4 positive charges per α β dimer to β 82 (EF$_6$) Lys a second chloride binding site as also envisaged by Arnone (1972). The best example in this regard was the description of the functional abnormalities of Hb Providence β 82 (EF$_6$) Lys → Asn or Asp by Bonaventura et al. (1976) and Manning et al. (1978) who demonstrated that this mutant Hb had a low oxygen

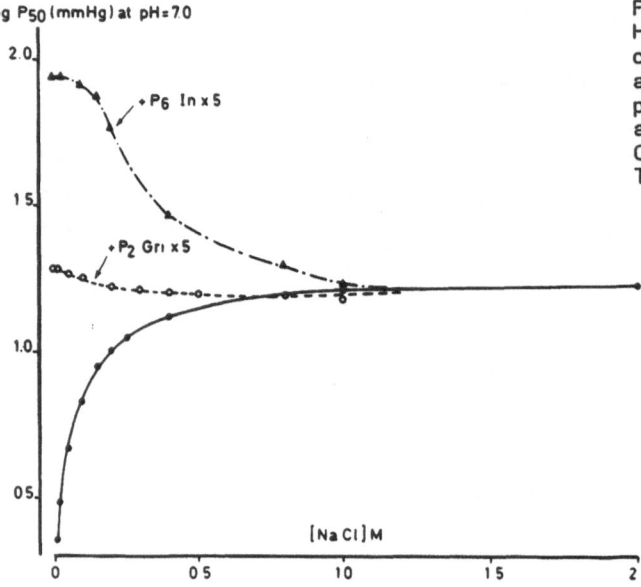

FIG. 6. Changes in oxygen affinity of HbAo (Log P 50, mmHg) upon increasing concentration of NaCl in the absence (filled circles) and in the presence of 5 mM DPG (open circles) and of 5 mM IHP (filled triangles). Conditions were: pH 7.0, 0.05 M bis-Tris, 0.6–0.7 mM Heme, T 25 °C.

affinity and both low DPG and low chloride binding and lowered alkaline Bohr effect. This was confirmed from proton NMR studies of Hb Providence which showed no downfield shift of the β heme resonances in the presence of chloride compared to HbA (Wiechelman et al., 1978). Two other mutants at the β82 (EF$_6$) site have been described, Hb Helsinki and Hb Rahere (Table 3) which both showed the same functional abnormalities in chloride and DPG binding. We have confirmed these results in a new case of Hb Providence Asn or Asp recently discovered in a french causasian family. Mutations at the other three DPG binding sites which have been studied in detail showed variable inhibition in DPG binding but normal interaction with chloride anion (Bonaventura and Bonaventura, 1978).

TABLE 3. Hemoglobin Variants at the β82 (EF$_6$) Lysyl Residue

		Oxygen affinity	n_{50}	Bohr effect	DPG	Cl$^-$
Hb Providence	β 82 Lys → Asn/Asp	↘	Nl	↘	↘	↘
Hb Helsinki	β 82 Lys → Met	↘	Nl	↘	↘	↘
Hb Rahere	β 82 Lys → Thr	↗	Nl	→	↘	↘
Hb X	β 82 Lys → Lys-Triose	↘	Nl	↘	↘	↘

Functional studies of various Hb variants at the $\beta_1 \beta_2$ interface have permitted to discover an important aspect of normal HbAo function. Taking Hb Providence Asn as an example, Bonaventura (1978) wondered why this mutant Hb had a lower oxygen affinity than HbAo in solutions of low chloride concentration. In Hb Prov. Asp the oxygen affinity is still further reduced compared to HbAo. They postulated that the 8 positive charges at the

$\beta_1\beta_2$ interface exert, in normal HbAo, a repulsive interaction which might destabilize the T deoxy structure and lower the allosteric constant L. If so the lowering of the density of these positive charges at the β cleft by susbtitution of one cationic group by a neutral one or an anionic group should raise the P_{50} and L and vice versa. This is confirmed in the former case for Hb Providence and Hb Helsinki. In the latter case, Hb variants where the substitution leads to a relative increase in the density of the cationic groups, have high oxygen affinities in the absence of anion as for example Hb Abruzzo ($\beta 143$ (H21) His → Arg), Hb Deer Lodge (β_2 (NA$_2$) His → Arg). The case of Hb Rahere (Table 3) which has a moderate increase in oxygen affinity has been tentatively explained by Perutz et al. (1980) by some specific steric effect of the threonine side chain. We have recently studied the functional properties of a minor component of HbAo discovered by isoelectric focusing electrophoresis in the hemolysate of a patient whose red cells lack completely 2,3 DPG but with 10 fold increase in trioses compounds. This is due to the absence of activity of the DPG mutase, the enzyme responsible for the synthesis of DPG (Rosa et al., 1978). Biochemical studies revealed that this minor component has a triose bound to $\beta 82$ (EF$_6$) Lys through a Schiff base linkage (HbX in Table 3). In all aspects, the functional properties of this abnormal post traductional modification of HbAo are similar to those observed in Hb Providence Asn or Hb Helsinki.

A further example of the specific role of single residue or group of residues at the origin of one of the functions of normal HbAo has been postulated, from the analyses of mutant Hbs by Perutz et al. (1980) who proposed that $\beta 143$ (H$_{21}$) His residue is involved, probably through chloride binding, for about half of the reversed acid Bohr effect (below pH 6.0).

V. MUTATIONS AT THE $\alpha_1\beta_1$ GLOBIN CHAINS CONTACT

In normal HbAo the $\alpha_1\beta_1$ contact is very tight and known to remain mainly unchanged in the oxy and deoxy configurations and not to participate in the allosteric properties, the alkaline Bohr effect nor anion binding of the molecule. Several mutant Hb have been discovered at residues involved in this contact leading as expected, and depending on the substitution, to an instability of the tetrameric and dimeric assembly of the molecule. Some of them are not instable and it is interesting to consider their functional properties. We have recently studied the oxygen binding properties of purified Hb San Diego which bears a substitution at the $\alpha_1\beta_1$ interface namely $\beta 109$ (G11) Val → Met (IHIC, 1984). Its stability is normal. We have succeeded to purify the abnormal component of this electrically silent Hb by the immobilized pH gradient electrophoresis technique (Rochette et al., 1984). Table 4 shows that despite a high oxygen affinity and decreased cooperativity, Hb San Diego retains normal heterotropic effects compared to HbAo both in the value of the alkaline Bohr effect and in anion binding. According to the proceeding discussion this is not surprising as the mutation in Hb San Diego does not concern any of the functional domain involved in the Bohr effect nor the anion binding sites.

TABLE 4. Heterotropic Effects on Oxygen Binding in Solutions of Hb San Diego and HbAo

	Hb San Diego	HbAo
Alkaline Bohr effect	- 0.487	- 0.53
DPG effect ($\Delta \log P_{50}$)	0.447	0.55
IHP effect ($\Delta \log P_{50}$)	1.014	1.097
P_{50}^{*} (mm Hg)	1.5	5.4
n_{50}	1.80	2.8

* Conditions were: pH 7.20, NaCl 0.1 M, bis-Tris 0.05 M, 25 °C

$\Delta \log P_{50} / \Delta pH$ (6.8-7.4)

DPG (Na salt) was 5 mM and IHP 2 mM

Heme concentration was 60 µM at pH 7.20, 25 °C

VI. CONCLUSIONS

Altogether these observations on Hb variants allow the general conclusion that the functional abnormalities due to the substitution of one residue involved in one specific functional property of normal Hb leaves the remaining function of the molecule unaltered. X-rays crystallographic analyses of mutant Hb have revealed also that the perturbation brought about by the mutation remains localized to the immediate neighborhood of the residues involved. This scheme applies rather well in mutant Hb located in the direct functional domain which have been considered. But it should be stressed that profound functional abnormalities may be observed in mutant Hb in which the substitution lies at distance from a critical functional residue. This is the case in mutant Hbs which imply an internal residue where the substituted side chain may disrupt the order of an α helix, create or disrupt hydrogen bonds and change indirectly the normal stereospecificity of the functional domain considered. This is observed for example in Hb Hope (IHIC, 1984) or in Hb St-Jacques (Rochette et al., 1984).

REFERENCES

Arnone A (1972) X-ray diffraction study of binding of 2,3 diphosphoglycerate to human haemoglobin. Nature (London) 237: 146-149

Arnone A, Williams DJ (1977) Crystallographic evidence for anion binding sites at the NH_2-termini of α subunits of human deoxy hemoglobin. In Molecular Interactions of Haemoglobin, INSERM Paris, pp. 15-22

Benesch RE, Benesch R (1974) The mechanism of interaction of red cell organic phosphates with hemoglobin. Adv. Protein Chem. 28: 211-237

Bonaventura J, Bonaventura C, Sullivan B, Ferruzzi G, Mc Curdy PR, Fox J, Moo-Penn WF (1978) Hemoglobin Providence. Functional consequences of two alterations of the 2,3 diphosphoglycerate binding site at position β 82. J. Biol. Chem. 251: 7563-7571

Bonaventura C, Bonaventura J (1978) Anionic control of hemoglobin function. In: Caughey WS (ed) Biochemical and Clinical Aspects of Hemoglobin Abnormalities. Academic Press, New York, pp. 647-661

Bunn HF, Forget BG, Ranney HM (1977) Human Hemoglobins. Saunders, WB, Philadelphia

Fermi, G, Perutz MF (1981) In: Philipps DC and Richards FM (eds) Atlas of Molecular Structures in Biology. 2. Haemoglobin and Myoglobin. Clarendon Press, Oxford, UK

Harano K, Harano T, Shibata S, Ueda S, Mori H, Seki M (1984) Hb Okasaki (β 93 (F$_9$) Cys → Arg) a new hemoglobin variant with increased oxygen affinity and instability. FEBS Lett. 173: 45-47

INTERNATIONAL HEMOGLOBIN INFORMATION CENTER (IHIC) (1984) Hemoglobin, 8: 243-300

Kilmartin JV, Wootton JF (1970) Inhibition of Bohr effect after removal of C-terminal histidines from haemoglobin β -chains. Nature (London) 228: 766-767

Kilmartin JV, Arnone A, Fogg J (1977) Specific modification of the α-chain C-terminal carboxyl group of hemoglobin by trypsin catalyzed hydrazinolysis. Biochemistry 16: 5393-5397

Kilmartin JV, Imai K, Jones RT, Faruqui AR, Fogg J, Baldwin JM (1978) Role of Bohr group salt bridges in cooperativity in hemoglobin. Biochim. Biophys. Acta 534: 15-25

Kilmartin JV, Fogg JH, Perutz MF (1980) Role of the C-terminal histidine in the alkaline Bohr effect of human hemoglobin. Biochemistry 19: 3189-3193

Manning JM, Nigen AM (1978) Major sites for the oxygen-linked binding of chloride to hemoglobin. In: Caughey WS (ed) Biochemical and Clinical Aspects of hemoglobin Abnormalities. Academic Press, New York, pp. 687-694

O'Donnell S, Mandaro R, Shuster T, Arnone A (1979) X-ray diffraction and solution studies of specifically carbamylated human hemoglobin Amer. J. Biol. Chem. 254: 12204-12208

Pettigrew DW, Romeo PH, Tsapis A, Thillet J, Smith ML, Turner BW, Ackers GK (1982) Probing the energetics of proteins through structural perturbation: sites of regulatory energy in human hemoglobin. Proc. Natl. Acad. Sci. USA 79: 1849-1853

Perutz MF, Lehmann H (1968) Molecular pathology of human haemoglobin. Nature 219: 902-909

Perutz MF (1970) Stereochemistry of cooperative effects in haemoglobin. Nature (London) 228: 726-739

Perutz MF, Imai K (1980) Regulation of oxygen affinity of mammalian haemoglobins. J. Mol. Biol. 136: 183-191

Perutz MF, Kilmartin JV, Nishikura K, Fogg JH, Butler PJG, Rollema HS (1980) Identification of residues contributing to the Bohr effect of human haemoglobin. J. Mol. Biol. 138: 649-670

Perutz MF, Brunori M (1982) Stereochemistry of cooperative effects in fish and amphibian haemoglobins. Nature (London) 299: 421-426

Perutz MF (1983) Species adaptation in a protein molecule. Mol. Biol. Evol. 1: 1-28

Poyart C, Bursaux E, Arnone A, Bonaventura J, Bonaventura C (1980) Structural and functional studies of hemoglobin Suresnes. J. Biol. Chem. 255: 9465-9473

Rochette J, Righetti PG, Bianchi Bosisio A, Vertongen A, Schnek G, Wacjman H (1984) Immobilized pH gradients and reversed-phase high performance liquid chromatography: a strategy for characterization of haemoglobin variants with electrophoretic mobility identical to that of HbA. J. Chromatogr. 285: 143-152

Rochette J, Varet B, Boissel JP, Clough K, Labie D, Wacjman H, Bohn B, Magne Ph, Poyart C (1984) Structure and function of Hb St-Jacques ($\alpha_2 \beta_2$ 140 (H18) Ala $\rightarrow$ Thr): a new high oxygen-affinity variant with altered bisphosphoglycerate binding. Biochim. Biophys. Acta, 785: 14-21

Rosa R, Prehu MO, Beuzard Y, Rosa J (1978) The first case of a complete deficiency of diphosphoglycerate mutase in human erythrocytes. J. Clin. Invest. 62: 907-915

Wiechelman KJ, Fox J, Mc Curdy PR, Ho C (1978) Proton nuclear resonance studies of Hemoglobin Providence. Biochemistry 17: 791-795

Index

172